TELECOMMUNICATION: EXPLORING COMPETITION

European Communication Policy Research Series

ISSN: 0926-9819

Telecommunication: Exploring Competition

edited by

Howard Williams and Mark Borman
University of Strathclyde
Glasgow
UK

1994
IOS Press
Amsterdam • Oxford • Washington DC

Ohmsha

Tokyo • Osaka • Kyoto

ISBN 90 5199 192 4 (IOS Press)
ISBN 4 274 90010 X (Ohmsha)
Library of Congress Catalog Card Number: 94-078816
ISSN 0926-9819

Publisher
IOS Press
Van Diemenstraat 94
1013 CN Amsterdam
Netherlands

Distributor in the UK and Ireland
IOS Press/Lavis Marketing
73 Lime Walk
Headington
Oxford OX3 7AD
England

Distributor in the USA and Canada
IOS Press, Inc.
P.O. Box 10558
Burke, VA 22009-0558
USA

Distributor in Japan
Ohmsha, Ltd.
3-1 Kanda Nishiki-Cho
Chiyoda-Ku
Tokyo 101
Japan

PRINTED IN THE NETHERLANDS

Preface and acknowledgements

The 8th European Communications Policy Research conference was held at Thainstone, Aberdeenshire in October 1993. Like previous conferences, this meeting had the aim of bringing together researchers, regulators, operators and users from the telecommunications sector, to discuss current issues concerning the development of policy in Europe. In the tradition of CPR, a series of refereed papers was presented and substantial time was devoted to discussion. This volume compromises the revised papers from the conference incorporating the key elements emanating from the discussions.

The papers were refereed by the international scientific committee whose sustained support and commitment to the CPR has ensured its position as one of the key meeting points of those concerned with the future regulation of the communications sector in Europe.

The conference was generously supported by the programme on Information and Communications Technologies (PICT) of the Economic and Social Research Council, Scottish Enterprise, Grampian Enterprise Limited, and the Network and Resource Management Centre at the University of Strathclyde, Glasgow.

The success of the conference and production of this volume have been secured by the support of three people. My joint editor Mark Borman assumed much of the responsibility for the day to day administration prior to the actual conference and his help during this period was indispensable. I am also indebted to Kina Rayner whose eye for detail and support ensured the smooth running of the conference. Finally, I must fully acknowledge the work of Dr Derek Manson-Smith whose skills as a technical editor have ensured the production and layout of this volume.

Howard Williams
July 1994

Introduction

The policy rhetoric of competition that emerged in the early 1980s has been sustained and continues to be enshrined in new legislation. In parallel with the introduction of new policy regimes centred on the tenets of liberalisation, regulation, deregulation, privatisation and competition has been the emergence of new policy questions, in particular the exploration of the consequences of the new policy regimes. Interwoven with this new policy framework has been the transformation of the technical infrastructures, which in itself has created sets of new policy questions not only about innovation and the diffusion of the technologies but also about the inter-relationship between new technologies and, for example, questions over privacy. It is in exploring some of these issues that the papers in this volume cohere.

What is evident is that across Europe a diversity of policy regimes exist and each has something to offer in exploring not only the limits to the current policy framework but also the detail of how the current regime manifests itself. The purpose of this volume is to bring together some of these different perspectives and draw out the implications for policy development. In so doing, the papers in this volume focus on four different dimensions of the current policy debate, namely, the local loop, new services innovation, consumer rights, and the issues surrounding liberalisation in the smaller countries in Europe.

The development of the local loop in a competitive environment is seen as one of the more complex policy areas, in particular the balance that needs to be created between the incentives to encourage alternative investment in the form of, say, cable-television networks, and the rights of the incumbent (and dominant carrier) to provide a full range of services. In their paper, Cave and Sharma review the situation in the UK, post the duopoly review. In essence, their papers charts the debate over the development of the local loop and its costs in the UK, where the basic policy assumption is one of end-to-end facilities-based competition. Their central contention is that British Telecom faces increasing competitive challenges to its dominance in the local loop as new entrants in both mobile services and cable-television networks continue to invest in new facilities. They, however, also comment on the way in which new technical developments, particularly video-on-demand, is creating some degree of regulatory uncertainty. In her paper, Thea Weijers charts the development of local cable-television networks in the Netherlands. In contrast to the UK, the development of cable television in the Netherlands can be seen as the result of collaborative ventures between local authorities and the cable operators. In examining whether or not cable-television networks are the missing link, her key argument is that the existing cable operators need to develop new interactive services to counter the increasingly competitive stances being adopted by the dominant carrier and satellite-television broadcasters.

The interplay between new regulatory regimes and new technologies has brought into question the nature of service development and the provision of new networks, and nowhere is this interplay as well demonstrated as with integrated services digital networks (ISDN). For some, ISDN is a force for integration and, therefore, in some respects, a throwback to the days of PTTs, for others it is a basic platform from which service innovation and customer-centred services can be developed. What is evident however, is that ISDN remains a set of services and technologies which continue to be surrounded by uncertainties, for example, in their uptake and use by customers. Three papers focus on the diffusion and use of advanced services, including value-added network services. In his paper, Stöckler reviews the diffusion process and highlights the importance of intermediaries and new marketing in promoting the innovative use of ISDN services. The unexpected success of ISDN in recent years is explored in the paper by Fuchs. He highlights the relatively poverty of some of the previous explanations of the use of ISDN, and through the use of comparative analysis across countries both within the European Union and elsewhere, points up the significance of institutional actors – in the case of Europe, of the European Commission. Finally, in this exploration of the uptake and use of advanced services, Henten and Ottosen examine the case of electronic-data interchange. Their conclusion is that electronic-data interchange does not necessarily need the support of value-added intermediary suppliers and that ISDN linkages may be far more effective and relatively cheaper.

An area of policy debate which is highly differentiated across Europe is that of consumer protection and the rights of consumers in a newly liberalised and competitive environment. It appears that in some countries questions over privacy and data protection are non-issues, whilst in others considerable controversy surrounds such issues, particularly in the context of an increasingly de-regulated environment. The argument advance by Kubicek is that the narrow focus of telecommunications regulation on economic issues, such as tariffs, is effectively preventing other policy debates, such as privacy, from surfacing. He observes that this economic perspective on regulation is being reinforced by European legislation on telecommunications and comments that it is only in a relatively restricted area of privacy that a draft directive has been submitted. His paper provides a comprehensive annex which illustrates the issues to be regulated, as well as the regulatory approaches already adopted by some member states. Adopting another stance, Arlandis critically reviews the concept of universal service and argues that its meaning and policy consequences can only be fully understood alongside an understanding of competition and the nature of industrial dynamics. His key point is that the definition of universal service alongside monopoly provision is relatively simplistic and that the inter-relationships between competition and universal service need to be more fully explored.

The final two chapters seek to explore the relevance and limits of the models of competition being enshrined in European legislation in the context of smaller and peripheral economies. In his paper, Caloghirou examines the development of Greek telecommunications policy during a period of political uncertainty in which, as a result of a general election, there was a change of government. His paper summarises the development of policy and provides a useful framework within which to understand contemporary changes in the telecommunications sector in Greece. In contrast, Preston examines the tradition and influence of political economists in shaping policy debates, drawing inspiration from the work of Cairnes. In particular, Preston's paper critically examines the extent to which the neo-liberal model and approaches to telecommunications regulation may be regarded as 'universal' in their reach and applicability. His conclusion is that the neo-liberal policy regime being promulgated by the European Commission fails to take account of the specific and special characteristics of the telecommunications sector in smaller and less-developed countries.

In summary, these papers sought to explore the complexities of competition as it is unfolding within Europe and to highlight some of the issues on which policy makers should focus their attention. In so doing, this volume begins to make a contribution, in the tradition of CPR, to the policy-making process in Europe by bringing together the richness and diversity of different perspectives. This volume is part of the critical foundations which seek to ensure that the benefits of competition are secured and enjoyed by consumers across Europe.

Contents

Telecommunication: Exploring Competition
H. Williams and M. Borman (Eds.)
IOS Press, 1994

Chapter 1

Competitive developments in local telecommunications in the UK

Martin Cave and Yogesh Sharma
Department of Economics, Brunel University, United Kingdom

Abstract: The market for local and long-distance telecommunications services has witnessed significant competitive entry, following the end of the duopoly review in 1991. Competition among mobile operators has also intensified with the introduction of new digital mobile networks. Allowing cable operators the right to supply voice telephony in conjunction with entertainment, proved to be an intriguing outcome of the review process. Developments in compression technology continue to harness the transmission capabilities of existing networks, so bringing forth opportunities for telecommunications companies entering new markets, such as switched video. This sets the background against which British Telecommunications will face a future challenge to its current dominance in the local loop.

1. Introduction

Competition in UK telecommunications has made considerable progress since the termination of the duopoly policy in 1991. Thus cable television operators, allowed to enter the market since 1991, now provide voice-telephony services to nearly 400,000 cable subscribers. The Department of Trade and Industry (DTI) in association with Oftel, the industry regulator, has already issued about 30 new licences. Some of these licensees intend to operate as long distance carriers: they include Energis – a subsidiary of the UK National Grid Company and other utilities. Other potential operators are offering or intend to offer local communications services using either wireless or wire-based technologies.

The UK telecommunications sector has also been the target of considerable amounts of foreign direct investment, most notably from the North American regional Bell operating companies (RBOCs). According to an estimate made by the British Cable Television Association, current investments amount to nearly £1 billion annually as cable franchises are busily constructing their networks.

The aim of this paper is to describe developments in competition in the local loop, to analyse likely costs of entry, and to examine regulatory responses. The next two sections describe the new technologies and present evidence on their costs. This is followed by an account of the regulatory framework. The final section examines how the competitive market is likely to develop.

2. Local delivery technology

The past few years have seen important reductions in the costs and potential service capabilities and quality levels of both copper and fibre in the local loop. Greater service possibilities from optical fibre have been developed for so-called narrowband and broadband networks. Improvements in information-transfer rates resulting from compression, have enhanced existing levels of bandwidth made available from cable installations (for example, the transmission-rate limit for metallic twisted pair is at least 4 Mbit/s over 100 metres, reduced to 800 kbit/s over an average line length of line). These and other developments support the transmission of low-definition television signals over the traditional metallic twisted pair. This radically alters the possibilities of the use of traditional technologies to offer new services, such as, video-on-demand in conjunction with standard voice telephony.

Although these technical improvements tend to alter the balance of advantage in favour of fibre over copper, British Telecommunications' (BT) current policy is based upon the maintenance of copper in large parts of the access network for residential subscribers, and its progressive replacement by fibre for large for larger business customers.

A debate now hinges on whether telephone companies like BT should be allowed to offer video-on-demand in direct competition with cable companies. It is conceivable that such services would be offered in conjunction with the development of integrated broadband networks (IBNs) offering video, voice and data on a single optical fibre. Alternatively, delivery on copper may be possible.

The feasibility of deploying IBN networks ultimately rests on the comparative cost of services being offered collectively via integrated means, or separately via narrowband and cable networks. In the US, Johnson and Reed [1] argue that the only innovative feature of telephone entry into cable television would be to offer residential subscribers a switched 'on-demand' video service. This contrasts with the simultaneously-delivered distributive video services, or 'near-demand' (since films tend to be repeated), currently offered by the cable companies. Telephone subscribers would be able to access a centralised video library covering a wide choice of entertainment needs. The idea of offering a 'switched' video service is analogous to telephone users choosing to switch to other telephone users. However, as is noted below, the key difference between the UK and the US is that in the former the cable network has not yet been built.

UK cable operators now have freedom to provide telecommunications services on their own account. They have the option of either running a separate telecommunications cable down the same duct as that used for the delivery of entertainment services, or of seeking for electronic integration between the two services. Although the latter is technically possible, the former is currently preferred. This is not surprising as cable companies are typically aiming for both residential and business customers of telecommunications services, while entertainment services are purchased almost exclusively by households.

A further wire-based local-access technology now available, is the construction of a technologically-advanced fibre-optic network through urban areas and business parks, known in the USA as competitive-access providers (CAPs). These networks now run through the overwhelming majority of the largest metropolitan service areas in the US, offering a service highly competitive with that provided by the RBOCs. Although such networks do not yet operate in the UK, roughly the same elements can be found in a licence recently issued to COLT, a company proposing to provide a fibre-based network within the London metropolitan area. Yorkshire Electricity and Kingston Communications have also recently announced plans to build a regional fibre-optic network spanning Yorkshire and Humberside, through a joint venture known as Torch Communications. Torch Communications along with Energis are aiming to offer long-distance services and wholesale re-selling to business customers in addition to selling other services to other telecommunications operators.

The more radical recent departure from traditional local-access technology involves the use of wireless communications. Since the two cellular networks started operation in 1985, they have obtained nearly two million subscribers. The coverage of their networks extends to 98 per cent of the UK population. The two cellular networks were augmented in the late 1980s by networks constructed by four telepoint licensees offering a portable (but not mobile) service: subscribers could make outgoing calls only if located within a specified range of a base station. This service initially failed, although it has recently been revived.

The more interesting recent development is that of personal communications networks

(PCNs), which offer mobility and two-way calling from any location. PCN technology uses digital radio operating at double the frequency of cellular radio, implying improved call quality and digitally-encrypted calls. Mercury's PCN system called 'One-2-One' was recently launched to customers within the London metropolitan area. The second operator – Hutchison Microtel – launched in 1994 a new service (Orange) with a larger coverage. Mercury's One-2-One service is planned to cover most of the UK population by 1996. Further digital cellular networks built by Vodafone have also begun to offer two new service, MetroDigital and EuroDigital, aimed at offering city-wide and European connection respectively.

The final radio-based technology considered here is non-mobile radio. Notable examples of this are the technologies to be employed under the licences granted by the DTI to Ionica and Millicom. These networks can interconnect with national trunk operators. Calls will be routed to base stations from the trunk network, at which point radio takes over. Point-to-point radio systems in the 1.3 to 4.5 GHz band achieve the final connection to end-users as signals are conveyed to roof-top antennae units. The link from the roof-top to the hand-set is essentially the same as for existing fixed services.

Table 1. Local delivery systems: technologies and services

	Distributive services					**Switched services**				
Transmission technology	**Low speed data**	**High speed data**	**PAL quality video**		**HDTV**	**Low speed data**	**High speed data**	**Voice**	**Low quality video**	**Video phone**
A – Wire-based										
Copper wire						/		/	/	
Coaxial cable	/		/							
ISDN over copper pairs						/		/	/	
HDSL over copper pairs						/	/	/	/	/
Fibre-to-the-curb (B-ISDN)	/	/	/		/	/	/	/	/	/
Fibre/coaxial cable hybrids		/	/	/						
B – Wireless										
Millimetric radio							/	/		/
Paging	/					/		/		
Cellular telephony	/					/		/		
PCS	/	/				/	/	/	/	
MMDS			/	/				/		
Cellular TV	/		/	/						
DBS		/	/	/						

Source: C. Berben [2]

All these technologies – wire-based and radio-based – have different potentials for providing distributive and switched services. Table 1 (adapted from Berben) [2] provides a useful summary. Clearly, different customers have different needs, and it is to be expected that markets will become segmented.

Telecommunications operators like BT are aiming to manage different traffic demands on its access network by employing a mix of new delivery systems including:

- point-to-point fibre systems (or active star network);
- passive optical networks (PONs) – (passive star network);
- high-speed digital subscriber loop (HDSL) which provides up to 1.5 Mbit/s over two pairs of loop-grade copper at distances of up to 3.6 km;
- radio.

These transport facilities are to be progressively deployed over the next five years and are the key to serving its major customers, the top 200 of whom have 90,000 UK sites each using more than one line. It is predicted that by 1997, 1000 sites will require access capacities exceeding 34 Mbit/s compared to 80,000 sites requiring 2 Mbit/s (Allen and Marshall) [3].

The development of the delivery modes noted above, is a powerful illustration of how the relaxation of regulatory restrictions can, in a comparatively brief period, elicit new technical alternatives. But it is important to recognise, even before a discussion of the costs of alternative local-access technologies, that the alternatives offer different types of service. These will clearly affect the willingness of different groups of customers to substitute one for another.

3. The costs and market potential of alternative-delivery technologies [4]

There are broadly three methods of investigating the costs of telecommunications technologies. The first involves the estimation of cost functions from a cross-section or time-series of observations. This approach achieved considerable popularity at the time of the break up of the Bell system, when rival econometricians produced estimates proving or disproving the existence of economies of scale and scope. Much subsequent work has focused on properties of the functions of the local exchange carrier. For example, Shin and Ying [5] undertook tests on a pooled cross-sectional sample of 58 local exchange companies (LECs) from 1976 to 1983. They concluded that the cost function is definitely not sub-additive, suggesting that the benefits in breaking up the monopoly outputs of existing LECs substantially outweigh the potential inefficiency. Unfortunately, the appropriate data for the UK telecommunications systems are available only to BT, which has probably not undertaken and certainly not published the results of a similar analysis.

The second approach to investigating telecommunications costs starts from the operators' management-accounting data. Thus, in jurisdictions where cost allocations are submitted to the regulator and placed in the public domain, it is possible to identify costs which can be attributed to particular services. For this process to work, the investigator must have access either to the raw data or have a detailed knowledge of cost-allocation procedures, for example, through the publication of a cost-allocation manual, which the operator must abide by in presenting data. The simple publication of fully-allocated cost data is not enough, as they are subject to the vagaries of the overhead-allocation system. Unfortunately, in the United Kingdom the only data available on the cost attributed to BT's various services are contained in its *Financial Results by Service* (FRBS) statements for 1988/89 to 1990/91, published by Oftel in January 1992. But it is impossible for investigators to strip out the overhead component from the reported results and thus derive an indication of causally-attributed costs.

The third method of investigating costs is to construct an engineering-economic model of alternative-telecommunications systems. An exercise of this kind was undertaken by consultants to Oftel and the DTI as part of the duopoly review (see Bell and Cave) [6]. An American study by Reed [7] attempts to model and cost a variety of network architectures for the independent and integrated-services delivery of narrowband, distributive video and switched on-demand video services by representative telephone and cable companies.

For our purposes, the principal conclusions of the UK studies are as follows:

(a) the costs of a wire-based telecommunications network built from scratch are heavily skewed towards access-related costs. In particular, trench and duct costs account for between 25% and 40% of total annualised costs per line, depending upon location. As a first approximation, the cost function comprises a large fixed element common to all homes passed, a component reflecting the connection of particular subscribers to the local loop and a relatively small additional usage cost, much of it associated with switching;

(b) in the case of a cable system, whether telecommunications and entertainment services are electronically integrated or whether separate wires are employed, a high proportion of systems costs are common to the two services. As a consequence, the incremental cost of the supply of telecommunications services by an existing cable operator is substantially less than the stand-alone cost;

(c) PCN and other radio-based technologies exhibit a different structure, with a smaller proportion of access-related costs and a higher proportion of usage-related costs. The evidence of the consultancy work undertaken for Oftel and the DTI suggested that PCN exhibits substantial cost advantages over fixed-link alternatives, irrespective of population density. However, some caution is appropriate in interpreting these figures, because at the time PCN was not yet fully developed. PCN (but not the non-mobile radio systems such as that proposed by Ionica) also imposes additional costs on the subscriber in terms of the acquisition of a more sophisticated handset;

(d) the costs of radio-based systems depend, in important respects, on the trade-off chosen between the number of base stations, the weight and cost of the handset and the quantity and nature of the spectrum employed. If spectrum availability and the size of the handset are not constraints, only a small number of base stations are required. As spectrum becomes scarcer, more base stations are required but the receiving instrument may also be smaller.

Table 2.

Network options	**Capital cost per subscriber ($)**	**Annualised operating cost ($)**
PCN network (1)	703	546
Existing infrastructure		
Telephone network (2)	780	178
Cable-television network (3)	424	137
Cellular network (4)	593	400

(1) Cost includes handset in a new residential area assuming 25 MHz allocation size and 10% household penetration.
(2) Per home passed. Network assumes a fibre-to-the-curb architecture.
(3) The cost of building a new fibre-based cable-television network at 60% household penetration.
(4) Assumes a handset cost of $300.

The studies have now been supplemented by further work (in the US) on the costs of PCN and of fibre to the home. Reed [8] examines the costs of building an independent PCN network, compared with those of other networks or of delivering PCN services in conjunction with existing telephone, cellular and cable-television networks. The market penetration of PCN operators depends in part on the availability and price of the spectrum, but more importantly on the degree to which existing networks meet the delivery needs of PCN wireless-telecommunications services. Table 2 summarises the estimated cost of the construction and operation of a stand-alone PCN network, and of a variety of other delivery systems. Reed also estimates that substantial economies of scope exist when PCN is operated in conjunction with an existing network.

Table 3. Investment cost ($) of separate networks per home passed (assuming 60% video penetration for 1.5 hours of video-on-demand per week)

Integrated broadband	1760
Narrowband	568
Cable television	368

Source Johnson and Reed [1]

Delivery of new services, such as video-on-demand, requires the construction either of a separate narrowband network, or of an integrated broadband fibre network. Johnson and Reed [1] estimate the comparative cost of these alternatives per home passed. Table 3 presents some of their results, which are based on constructing completely new networks for 25 groups of 1024 households, each served by its own remote distribution units (RDUs) in the centre of the service area (25,600 households in total).

These figures suggest a strong case for building a separate narrowband network for telephony and switched on-demand video services, since the combined cost of narrowband and cable television (for distributive video only) is significantly cheaper than that of an IBN. The higher investment cost of the IBN is due to the greater need to install expensive electronic equipment. The electronics in a fibre network also tend to be more decentralised, compared with traditional copper telecommunications networks because fibre conducts insufficient amounts of electricity to power customer-premises equipment. An IBN network therefore incurs higher operating and maintenance costs.

Johnson and Reed also calculate annualised and incremental figures for providing a switched video service on a per video-subscriber basis, assuming a 60% response rate for this service. The figures in Table 4 assume different rates for the capital costs of a narrowband network (13%) and broadband network (17%).

The figures illustrated in Table 4 suggest a crude measure of calculating the incremental cost of providing a switched video investment per home passed ($824) as well as an total ($1375) and annualised ($297) investment cost per video subscriber. On the assumption that O&M expenses for on-demand video service is 15% of the incremental investment of on-demand video service ($1373), the total annualised operating cost per video-subscriber comes to $503. Table 5 presents a more complete set of result for alternative levels of penetration and use of video services.

Table 4. Annualised incremental investment cost ($) of video-on-demand (assuming 60% video penetration for 1.5 hours of video-on-demand per week)

IBN investment	1760
Less investment cost of narrowband network	568
Less investment cost of cable-television network	368
Leaves switched video incremental investment per home passed	824
Switched video incremental investment per video subscriber	1373
Switched video annualised investment per video subscriber	297

Source Johnson and Reed [1]

Table 5. Incremental annual cost ($) of on-demand video service per subscriber

% On-demand video penetration	Hours of on-demand video per week per subscriber 0.75	1.5	3	6
40	694	772	746	784
60	481	503	530	562
80	372	393	419	448

Source: Johnson and Reed [1]

Since US households currently pay $300–350 ($380 for UK households) annually for all cable services, the estimated annual cost of on-demand video appears to be prohibitive at the outset. An IBN network will typically require video switching, multiplexing and expensive electronics, while alternatives to 'on-demand' video exist in the form of 'near-demand' services supplied by cable companies and video-rental outlets offering a bank of pre-recorded entertainment.

These results have aroused some debate (Goodman et al [9]; Johnson and Reed [10]). But the development of compression techniques permitting the use of existing copper plant to provide video service, further undermines the viability of the integrated network. Unfortunately, the relevant technical and cost parameters of this delivery mode are not yet fully understood, and we can present no estimates of them.

Thus, the final penetration of new transmission technologies such as optical fibre remains an empirical question rather than a foregone conclusion – as assumed by many commentators. If technical developments continue to harness the potential of existing network architectures, the availability of new services may not be trapped under the weight of network 'bottlenecks', especially if the mix of services offered to users coincides with their demand for new services.

In summarising this discussion of network costs, we draw attention to three points in particular. The costs of a new wire-based network are heavily skewed towards access with usage costs low and diminishing. However, many of the access-related costs of an existing network are sunk or non-salvageable, freeing the operation to make competitive responses. Second, radio-based systems exhibit a different cost structure, even though there is continuing debate about the level of costs which they impose. Third, the costs of new interactive services such as video-on-demand for the UK are as yet unknown.

4. Regulation and competition in the local loop [4]

Although the abandonment of the duopoly policy liberalised entry into telecommunications-service markets, many important restrictions have remained. In particular, BT is prohibited from conveying or providing broadcast-entertainment services until a review which – according to the Government policy statement – will not occur until 1998.

This regulatory policy differs from that followed in the United States, where the Federal Communications Commission (FCC) has rapidly been reducing the restrictions on participation by telephone companies and cable companies in services traditionally provided by the other. The key difference between the two countries, of course, is that in the United States, 90% of all homes are passed by cable, while the equivalent figure in the United Kingdom is approximately 15%. There is thus a clear risk, even if it is exaggerated by UK cable operators, that allowing BT into the provision of entertainment services would interfere with the construction of the rival cable network. Since both networks are 'sunk' in the United States, the FCC formed the view that promoting competition between them would benefit the public through lower prices and the encouragement of new services. Moreover, each industry has its own strength – cable companies in the provision of entertainment and telephone companies in the routing of two-way information flows. Liberalising entry conditions is thus likely to lead to competition based upon product differentiation.

However, there are now serious doubts about whether BT is in fact restricted from providing video-on-demand. The crux here is the definition of a broadcast service. If it is defined in point to multipoint terms, then an arrangement such as video-on-demand, under which each household views programmes from the same video library, but on a non-simultaneous basis, does not qualify as broadcasting. The Independent Television Commission (ITC) has now acknowledged that video-on-demand does not require a local broadcast-delivery licence, although further legal challenge is possible (*Financial Times*, 30 September 1993, p 20).

As far as other aspects of structural regulation are concerned, as we noted in the introduction the new arrangements followed the end of the duopoly policy. The number of mobile operators of each type is determined through a regulatory decision-taking process. Cable-television operators may apply for a licence for telecommunications services through a straightforward procedure. And any applicant may apply for a fixed-link telecommunications-operating licence. Although such applications are considered 'on their merits', severe criteria are not applied.

Regulation of conduct is determined above all by BT's licence and the associated price-control regime. The licence imposes service obligations and puts bounds on price discrimination. In particular, any quantity discounts have to satisfy a complex set of conditions, even though in the longer term incremental cost pricing will be permitted.

As far as the overall level of BT's prices is concerned, the key element of price control is a price cap linked to the Retail Price Index (RPI). The severity of this price cap has increased from RPI –3 (1984–89), to RPI –4½ (1989–91), to RPI –6¼ (1991–93) and RPI –7½ (1993–97). The coverage of the basket has also increased over the relevant period, notably by the inclusion of international calls in 1991. BT may not count for the purposes of satisfying its price cap, any further quantity discounts offered in addition to those available before August 1993.

The RPI –X price cap for the tariff basket as a whole is augmented by side constraints on the prices of particular services. Since 1984, BT has been restricted from increasing its quarterly rentals by more than 2% above the RPI. As the overall cap has tightened, this restriction has permitted increasing and substantial rebalancing between access and usage charges. Nonetheless, the company asserts that developments in technology are making a progressively higher proportion of costs access-related rather than usage-related, and that in consequence the losses the company is experiencing on its access business are increasing.

The company demonstrates this proposition by the *Financial Results by Service* data discussed above. A summary of the results for 1991–92 is shown at Table 6.

Table 6. BT's Financial Results by Service 1991/92 (£m)

	Access	Inland* calls	International calls	Directory enquiry calls
Turnover	2382	4765	1588	179
Return	–732	2871	682	–28
Capital employed	6398	4959	829	107
Return on capital employed (%)	–11.4	57.9	82.3	–26.2

* local, national and interconnect

These calculations are subject to the standard difficulties associated with the fully-allocated costs – notably the element of arbitrariness in allocating overhead costs among different services. Nonetheless, as they stand they demonstrate a substantial negative return on access (of –10.5% for business access and –11.8% for residential access). Assuming a capital cost of 15%, this represents a loss of £1.7 billion. In 1990/91 (for which fuller data were available), BT's rate of return on its access, business and local combined was 5%.

In response to BT's representations about the access deficit shown by the FRBS, Oftel agreed in 1991 that it should be shared with other operators interconnecting with BT, through the payment by the latter of access-deficit contributions (ADCs). These are calculated proportionately to the profitability of the service sold by the interconnecting operator. Thus, a competing operator supplying a minute's international call would pay to BT a larger ADC than would be paid by the same operator supplying a customer with a minute's local call. However the Director General of Telecommunications (DGT) has the power to waive in whole or in part the ADC payable by a 'small' operator. An operator supplying less than 10% of the relevant market may have the ADC waived, provided that BT's share of the relevant market is not less than 85%. The definition of the relevant markets is made by the DGT. Waivers cease in 1997, except for PCN operators, which may retain them.

The final major element of the regulatory environment relevant to the development of local competition concerns the principle upon which interconnection charges are established. Under BT's licence condition, these are established on a fully-allocated cost basis including the attribution of appropriate overheads and the use of an appropriate cost of capital. Oftel is currently redetermining the interconnection charges paid on a reciprocal basis by BT and Mercury, and an announcement is expected shortly.

This complex set of arrangements governing interconnection is subject to much current debate. In June 1992, the DGT issued a statement expressing the view that accounting separation between different BT businesses was necessary if competition was to develop and if there was to be public confidence that BT was not abusing its dominant position. Oftel asserted that accounting separation would achieve the objectives of transparency and non-discrimination: transparency through the exposure of costs relating to each of BT's businesses, and non-discrimination by ensuring that services offered by one BT business to another are offered on similar terms to non-BT businesses. In March 1994, Oftel published a statement indicating that it proposed to achieve these aims by requiring BT to publish separate retail and wholesale accounts for its activities [11].

With BT's current dominance of local telecommunications markets, the form of regulation of its conduct is critical to the development of competition. Given the company's dominance of the local market, the regulatory pricing of services creates an umbrella beneath which competing suppliers can or must shelter. To persuade customers to desert BT in their favour, competitors would normally have to offer a price advantage sufficient to match the cost of switching to an alternative supplier, which – given the absence of number portability – may at present be substantial. Given this fundamental background condition, Oftel is seeking to strike a balance between BT and its competitors in two ways: by imposing accounting separation upon the company and by requiring the sharing of BT's access deficit, although the latter is qualified by the use of waivers as an explicit entry-promoting device. These arrangements provide a complicated and shifting background against which competitors for local telecommunications services have to position themselves in the market.

5. Competition in the local loop

Official data on market shares in US telecommunications are not available, but a recent estimate of BT's involvement in various markets is given in Table 7.

Table 7. BT share of UK public switched telephony market – % of total UK market

	1984/5	1990/1	1991/2	1992/3
All business lines and calls	100	94	90	87
All residential lines and calls	100	99+	99	98
Inland calls	100	96	94	93
local	100	99	98	97
long distance	100	94	92	90
International calls	100	89	83	80
All telephone services in financial district of London	100	70	n/a	<50

Source: Attenborough [12]

The estimates show the accelerating success of entrants, especially in particular markets such as the City of London and international calls in general.

There is also a set of projections made by Oftel of BT's loss of market share to mid-1997. These were prepared as part of the review of BT's price control up to that year, and are given in Table 8 in two variants, a low and a high estimate.

Not surprisingly, these data project lower penetration by competitors in local markets. However, recent developments in wireless technology are encouraging for the development of price competition there. Over the summer of 1993 a pre-emptive response on the part of cellular operators to the emergence of PCN could be observed. Both network operators introduced a new tariff system designed to attract low users: such customers receive a reduction in their connection charge and monthly rental, but pay approximately twice the standard rate for call minutes. More radically, Mercury's tariff structure for its One-2-One PCN service involves offering free off-peak calls for all local calls, in addition to offering a unit subsidy of £150 for new (digital) technology handsets, matching a similar subsidy offered by the cellular-service providers. At the same time, its call charges will be lower than those for cellular operators. The availability of a variety of mobile services has also yielded a variety of tariffs to choose from. Table 9 lists the system of tariffs currently applicable to mobile services in the UK, including BT for comparative purposes. By taking a typical basket of calls as given, it is possible to compare the services, including a notional component for handsets. By bundling all costs together, Table 10 provides an annualised comparison of total costs, where each is composed of relevant connection, rental, handset and usage charges. (Handset and connection charges are spread over five years.) The calculations suggest that Mercury One-2-One is the cheapest radio-based service, but still far more costly than BT.

Table 8. Oftel's projection of BT's loss of market share

	Low estimate (%)	High estimate (%)
Access lines	5	14
Local calls	5	14
National (long-distance) calls	15	27
International calls	25	51
Implied rate of return on capital (%)	18.5	16.5

Source: Oftel [13]

Table 9. Mobile and fixed service costs and tariffs as at October 1993 (£s and pence)

	Vodafone				Mercury One-2-One		Cellnet			BT
	Prime time	Low call	Euro digtl	Metro digtl	Residen tial	Busi-ness	Prime time	City time	Life time	
Handset (£)	c100	c100	c400	c400	250–300	250–300	c100	c100	c100	c30
Connection (£)	58	29	59	59	24	24	59	59	29	116
Rental/month (£)	29	18	32	24	15	24	25	24	15	8
Peak tariff/ minute (p)	29	59	29	29	29	19	29@	24 59*	50	4
Off peak tariff/minute (p)	12	24	12	12	Free~	9#	12	12 12*	20	2

~ applies to local calls, 10 pence otherwise
* calls outside M25 motorway
\# all calls between One-2-One handsets are 8 pence
@ 33 pence for calls made in London

The rivalry described above, between PCN and cellular operators essentially involves firms using technologies with the same cost structures. What is more revolutionary, is the prospect of inter-technology competition between wireless and wire-based systems. The issue here is unusually complex. The normal expectation is that customers would segment according to the cost structures of the operators, which themselves would be reflected in their tariffs. Thus a wire-based technology, with substantial access costs and low usage costs, would be especially attractive to high users, while wireless technologies, with lower access but higher usage charges would attract low users. Even though it is unexpectedly difficult to demonstrate, using conventional economic models of competitive interaction, that this outcome would necessarily come about, the implied link between costs and tariffs is the likely outcome.

Table 10. Annualised operating costs for mobile services (£ per annum)

	Vodafone				Mercury One-2-One		Cellnet			BT
	Prime time	Low call	Euro digtl	Metro digtl	Residen tial	Business	Prime time	City time	Life time	
Handset	20	20	80	80	55	55	20	20	20	6
Connection	12	6	12	12	5	5	12	12	6	23
Rental/month	326	212	388	282	176	282	353	282	180	92
10 min peak call/month	35	71	35	35	35	23	35	28	60	13
10 min off peak call/month	14	28	14	14	–	11	14	14	24	4
Total cost	407	336	529	423	271	376	434	356	290	138
Without handset	387	316	449	343	216	321	414	336	270	132

However, the degree of regulatory constraint imposed on BT will play a key role in such competition. As noted above, the company will have increasing freedom to bring marginal prices down to incremental cost. Moreover, as many of BT's local-network costs are sunk, the company is likely to prefer price cutting to loss of market share. It can combine this with cost cutting, in a fast-expanding market characterised by economies of scale.

But despite its financial strength, BT is attacked from too many quarters simultaneously to be able to survive unscathed. Indeed, what seems to be happening in the local loop in the UK matches the vision of George Calhoun [14], which he described as a 'laminated access' or 'laminar network'. In Calhoun's words, this is a world in which 'multiple carriers provide competing, redundant (duplicative if you like) access facilities that, in effect, overlay one another. The kinds and qualities of the access offered are not necessarily identical.... In short, a laminar network is a series of partly competing, partly complementary, somewhat differentiated, overlapping access fabrics.' He goes on to argue that all competitive markets are laminar, as they all involve multiple overlapping distribution structures aimed at the same population base.

If this vision is correct, as it appears to be, the local loop may be losing its distinctive characteristic as a monopoly. We will soon see a multiplicity of competing networks using a variety of technologies and employing quite different tariff structures. Many customers will be connected to several networks. Interconnection among them will be ever more important, but that too stands a chance of becoming competitive. Such a development will change the face not only of the market place but also of the regulatory system. The era of regulation through competition may be upon us faster than we expected.

Notes and references

[1] L. L. Johnson and D. P. Reed, Telephone company entry into cable television: an evaluation, *Telecommunications Policy*, 1992, pp 122–134.

[2] C. Berben, Access to the Local Loop: perspectives for liberalisation in the EC. Paper presented at the 5th Economist Telecommunications Conference, Vienna, 23–24 September 1993.

[3] P. Allen and J. Marshall, Integrated Services Delivery to Major Customer Sites. Working Paper, 1993.

[4] This section draws in part upon M. Cave and Y. Sharma, Foreign Entry and Competition for Local Telecommunications in the UK After the Duopoly Review. In E. Bohlin et al (Eds), The Race for European Eminence in Telecommunications. Elsevier Science Publishers BV, Amsterdam, 1994.

[5] R. T. Shin and J. S. Ying, Unnatural Monopolies: local telephone, *Rand Journal of Economics*, 1992, pp 171–183.

[6] A. Bell and M. Cave, Lessons from the UK Duopoly Review. In F. Klaver and P. Slaa (Eds), Telecommunication: new signposts to old roads. IOS Press, Amsterdam, 1992.

[7] D. P. Reed, Residential Fibre Optic Networks: an engineering and economic analysis. Artech House, Norwood, Maryland, 1991.

[8] D. P. Reed, Putting It All Together: the cost structure of personal communication services. Office of Policy and Plans, Working Paper No. 22, Federal Communications Commission, 1992.

[9] M. Goodman, K. Lu, W. Sharkey, P. Srinagesh and N. Stolleman, Telephone Company Entry into Cable Television: a re-evaluation. Bellcore Economics Discussion Paper No. 96, 1992.

[10 L. L. Johnson and D. P. Reed, Telephone Company Entry into Cable Television: a reply, *Telecommunications Policy*, 1993, pp 234–235.

[11] Oftel, Interconnection and Accounting Separation: the next steps, 1994.

[12] N. Attenborough, Regulation of Competitive Telecommunications Markets. NERA Topics, National Economic Research Associates, London, 1993.

[13] Oftel, Policy on Separation and Interconnection, June 1992.

[14] G. Calhoun, Wireless Access and the Local Telephone Network. Artech House, Norwood, Maryland, 1992.

Telecommunication: Exploring Competition
H. Williams and M. Borman (Eds.)
IOS Press, 1994

Chapter 2

Missing links in the evolution of Dutch cable-television networks

Thea Weijers
TNO Centre for Technology and Policy Studies, Apeldoorn, the Netherlands

Abstract: The Dutch cable networks have reached a high penetration rate, providing good quality at a reasonable price. Support from local government and the freedom to retransmit satellite television are key factors for this success. To continue the success and resist the competition from PTT Telecom and satellite television, cable operators need to provide new services, such as interactivity and premium channels. This requires cooperation between operators and service providers to unite the fragmented networks and supply customer-premises equipment. However, the national telecom policy, proposed by the Minister, does not support this kind of cooperation. On the contrary, it still favours PTT Telecom.

1. A success story?

About 92% of all homes in the Netherlands are passed by cable. Cable operators expect to increase that percentage to about 97% in the next few years. Also, about 92% of all homes passed have a subscription, which is much more than the 60% that American operators usually consider the hallmark of success. In total, about 86% of all Dutch households have a cable subscription [1]. Most cable networks offer about 15–25 channels for an average monthly subscription of 18 guilders or US$10. Total invested capital is 3.5–4 billion guilders and the annual turnover is 900 million guilders or £315 million [2].

What are the reasons for this success? One reason is that building the cable networks was (financially) supported by local municipal councils, who wanted to get rid of the aerial antennae that polluted the skylines of densely-populated cities. Initially, some people were obliged to get rid of their antenna and, therefore, had no choice but to take out a subscription. Councils also saw cable television (CATV) as part of their local media or culture policy.

Another reason was that CATV promised better quality and more channels than people could pick up from the air. At first, this meant offering public broadcast from neighbouring countries (Belgian BRT, German ARD, ZDF, West 3, in some cities the BBC). However, in the eighties when most CATV networks were built, (commercial) satellite television arrived and, once it was allowed in, it turned out to be a big boost for CATV. Finally, cable television really had more to offer than on-the-air transmission.

Despite the achievement, CATV cannot be called a simple success story. Some of the networks are old and operators find it difficult to invest in new technology. There are few experiments with new services and most networks are only used for the retransmission of television and radio channels. At the same time, the former companion, satellite television, has grown to such an extent that it might turn into a competitor.

Some of the problems CATV operators face are a result of its past, which is discussed in the next paragraph.

2. A short history of CATV in the Netherlands

CATV networks were prohibited from offering anything other than broadcast because the Dutch telecommunications company, PTT Telecom, had a monopoly or a 'first claim' on all telecommunications infrastructures on public ground. When municipalities decided to have CATV networks built, they had to apply for a special permit to deviate from the PTT monopoly. This permit was given, provided that (1) PTT Telecom did not want to build the CATV networks, (2) the CATV networks were within municipal limits, (3) the CATV network only offered broadcast signals, and (4) the CATV network offered all public broadcasting that was in Dutch (which they must carry) and all public broadcast signals that could be received by air in the region.

If a CATV operator wanted to do more than retransmit broadcast signals it had to ask permission from PTT Telecom and, after its privatisation in 1989, from the Ministry of Traffic and Public Works (Directorate of Telecommunications and Post).

The result was that in 1991, the Dutch cable world consisted of over 400 separate local CATV networks, most with their own head station to receive broadcast signals, with different technical infrastructures and management systems. There is no national infrastructure to connect the CATV networks. The networks are operated by about 180 operators [3].

In some cases the CATV operator owns the CATV network, and in others the network is owned by one party (for example, the municipality, a building association) and operated by another (for example, an independent body, the municipal-utility company, the regional-electricity supplier) [4]. In many cases the municipal council owns the special permit and uses this to control part of the content of the network, deciding which signals the network should offer to subscribers. In others, the channels are selected by a programme council, consisting of representatives of the different groups in a community.

In all cases the administration of the CATV networks is simple. The operator's income is based on a fixed contribution by the subscriber plus a relatively small income for the distribution of certain commercial programmes, such as RTL 4, minus the fixed costs of the network and head station and the variable costs of exploitation (personnel, copyright, payment to programme suppliers). The operator offers each subscriber the same service (an extended antenna service) for the same price. As mentioned above, the content of the service (the programme package) is seen as a matter of public choice. Many municipal councils had lengthy discussions on whether to include Eurosport, MTV or CNN.

At the time that most CATV networks were built, the political debate on telecommunications concentrated on integrated services digital networks (ISDN). A commission chaired by Professor Zegveld was set up to advise on the best way to introduce ISDN. In 1988, the commission suggested bringing together the local infrastructures of the PTT telephone network and the CATV networks in one broadband network under the control of PTT Telecom. Naturally the CATV operators, organised in the VECAI (association of cable operators and owners), did not really welcome this plan, but the ISDN concept was so compelling that they did start talks with PTT Telecom to see how this so called 'unity of control' could be brought about. This obviously dampened all possible desire to experiment with new services or invest in anything else but the building of the actual networks; an investment which could easily be recovered.

Recently however the 'Zegveld scenario' was abandoned and the future for the CATV networks has become more promising.

3. 'Times they are a'changing'

In March 1992, the Minister of Traffic and Public Works, who is responsible for telecommunications policy, announced that she intended to allow CATV networks to provide datacommunications and to connect their networks. This would mean that the CATV networks would be allowed to compete with PTT Telecom and develop into telecommunications operators (with the exception of voice telephony until 1998). In a recent policy statement of July 1993 however, the minister suggested a slightly different scenario. She wants to allow one competitor beside PTT Telecom in wired communications. She also expressed a strong preference for this competitor to be a consortium of operators, who already have permission to operate an infrastructure: the CATV operators, Dutch Railways (NS) and the Dutch electricity companies [5]. The latter two own and operate a national long-distance telecommunications infrastructure to support their own operations (data processing, signal processing, etc). These infrastructures are supposed to provide the national backbone of the network, with the CATV networks providing the local loop. The consortium would be allowed to provide transport capacity and services (again, except voice telephony). The minister expects that this scenario will be best suited to fulfil the needs of the business user (especially for leased lines and back-up facilities) without harming the position of the private user. The proposed scenario raises questions:

- Neither of the operators mentioned has experience with providing datacommunications to third parties. In many cases the CATV networks do not extend to business areas. The networks are not switched, which might create problems with confidential data, even when methods from research on passive optical networks are used. New investments will be required. So will this decision be sufficient to introduce real competition? Or is more support for the cooperating operators as new entrants necessary?
- Royal Dutch PTT (KPN – of which PTT Telecom is a subsidiary), is a major shareholder of one of the largest cable operators, Casema. KPN recently bought all the shares of another large shareholder, so that it now owns approximately 95% of all shares. Despite strong denials by PTT Telecom and Casema, many are afraid that Casema could function as a fifth column. PTT Telecom is building a broadband video net, BVN, to distribute broadcast signals and other services to the feeder loop of CATV networks. In June 1994, KPN announced that it will set up a subsidiary company of PTT Post and PTT Telecom, Interactive Media Services (IMS). KPN intends to coordinate its activities on new media and new services (home-shopping, video-on-demand, interactive games, etc) through this centre. If the consortium of CATV, Dutch Railways and the electricity companies is not allowed in telephony, should PTT Telecom not be banned from video services?

These questions have been raised, not only by experts but also in parliament [6]. But it seems safe to say that the minister and parliament are reluctant to do anything that might seriously hurt the position of PTT Telecom. Both minister and parliament want PTT Telecom to be internationally competitive.

Even if the cooperation between the CATV operators and the electricity companies on datacommunications is successful, this will still leave the CATV operators with some problems in the provision of (new) video services to private users. This is addressed in the next paragraph.

4. From distribution to individual selection: redefining CATV

The CATV networks are structured as distribution networks, using a so-called mini-star topology shown in Figure 1. Most CATV networks use coax, and in some cases fibre (sometimes leased from PTT Telecom) is used for the feeder loop. Because the networks were meant for distribution, creating interactivity is still a problem. In some networks all one-way amplifiers are replaced with two-way amplifiers and a channel is made available for the returning signal. Most networks however, cannot provide this kind of interactivity yet.

At present, most cable networks carry one pay-TV channel, Filmnet [7]. This channel provides its subscribers with decoders. The decoder is part of the Filmnet service, not of the cable network. The cable operators only transmit the scrambled signal, for which they receive a fee from Filmnet. The operators are not involved in the administration or the marketing of the pay-TV channel.

The commercial channels RTL 4 and RTL 5 (owned by the Luxembourg company, CTL) offer an interactive teletext service. The interactivity is provided by telephone. Again, the cable operators are not involved in the administration.

There are a few cable networks experimenting with interactive services, such as an interactive-teletext service. One operator (NKM, recently taken over by Casema) offers a connection to the inter-university network, SURF, for students in the town of Nijmegen. Several operators have started an alarm service for handicapped people and elderly people.

The introduction of new services, such as pay-TV, near video-on-demand, etc, is severely hindered by the fact that cable operators have no experience of offering services other than open distribution (Table 1). In some cases, operators have experimented with offering different packages of channels. In most cases a limited and an extended package was offered. The limited package was cheaper and meant for people who otherwise might be unable to take out a subscription. Most operators abandoned this strategy, because the cheaper package was more expensive to provide, since it meant investing in filters to stop the signals going to those subscribers who had taken out the limited package. It also turned out that it was not the poor but the affluent who took out the limited package. The packages were rarely seen as a way of providing a more individual selection, replacing the collective selection by the municipal council, but rather as an instrument for income-policy; and as such it failed.

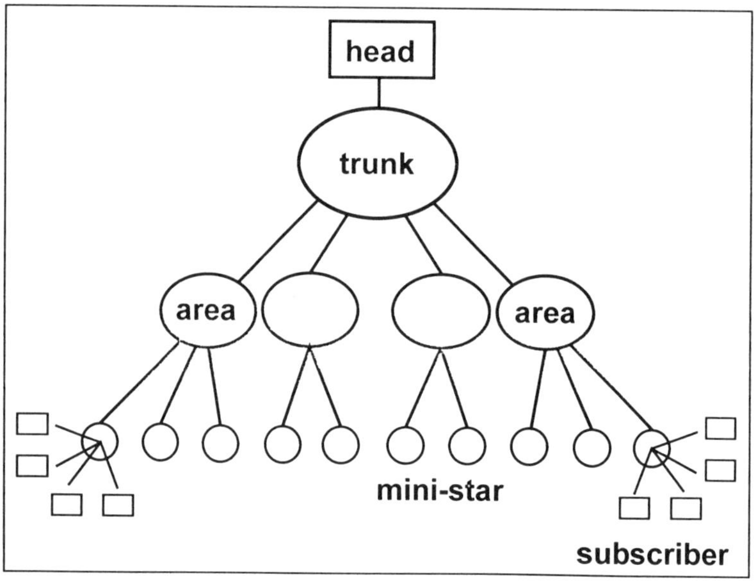

Figure 1: Cable network with mini-star architecture

Table 1. From service supplier to subscriber: functions in a cable network.

Service supplier

Infrastructure	**Intermediate services**	
	value added	**supporting service**
signal reception	open distribution	agreements with service suppliers
signal distribution	controlled access	customer service
signal registration	selected access	billing and collecting
point-to-point transmission	gateways to information services	distribution of terminals
		marketing

Subscriber

Both operators and subscribers tend to see cable as an extended antenna service. Subscribers often confuse the subscription to cable with the 'broadcast contribution' which every viewer or listener (that is, everybody who owns a TV or radio) has to pay to support the public-broadcasting system. Subscribers make no distinction between the cable network and broadcasting.

So far, most subscribers have not complained about the 'collective selection', but there are indications that this might change. There are far more satellite channels than any cable operator can distribute. Selection is no longer a matter of 'filling the cable channels' but involves making choices between (often commercial) channels. The municipal collective selection does not fit very well with a growing diversity or individualism in society. There are new standards for television and radio (HDTV, digital radio) that some subscribers may want and others not. In short, new methods for a more individual selection will have to be provided.

This means that the current concept of cable television as an 'antenna service' has to be redefined. Cable networks have to be presented as networks offering access to a number of value-added services, such as closed access (pay-TV, business television, digital radio), selected access (pay-per-view, video-on-demand) and information transport (videotext, gateways to databases, etc). The subscription to such a network is no longer for all the services provided, but for access to the network and open distribution only. Other services will be extra.

This means introducing terminals for decoding the signals and for the interactivity necessary for selected access and information transport. If the cable operators want to do more than distribution, it is vital that they have some control over these terminals; in fact these terminals should be seen as part of access to the network. As mentioned before, this is not the case at present: there are separate terminals for Filmnet; there are other service providers planning to offer a terminal, using the PTT Telecom network for interactivity; and there is a digital-radio station that provides its own digital tuner. It will not be possible for the cable operators to ban these terminals, but operators may try to provide their own terminals, offering interfaces to all services. This will have to be supported by an administration system for billing and marketing, that will be open to all service providers. This will lower the cost for the provision of new services (economies of scale). It will also make it possible for cable operators to generate more cashflow.

A group of large cable operators, ADEM, has started work on this terminal hoping to set a standard that will be acceptable to all operators and service providers. At the moment however, it is not clear whether such a strategy will be successful. Filmnet, RTL 4 and even PTT Telecom are also working on a terminal which can be used by others. It is also possible that certain facilities will be incorporated in television sets or computer terminals (multimedia), as has been the case with teletext facilities. The outcome of this standardisation process will be important for the position of cable operators as providers of value-added services. This, however, should not lead to a situation in which the standardisation process prevents experimentation and further development of the networks. The technologies involved are relatively new, and setting a fixed standard too early might jeopardize their development. Experimentation with different standards and interfaces will be necessary. This will require cooperation between some cable operators and some service providers. It is regrettable that the minister did not announce a policy to support this kind of cooperation, especially because recent signs show that ADEM's strategy is failing.

A point not yet addressed is the fragmentation of the networks. This has led to high fixed costs, because each networks has its own head end. It has also made it more difficult for new services that depend on national distribution to find a niche market. Standardisation might support these services but integration of the networks will also be necessary. This does not mean that local operators will have to hand over their networks to one central operator, but it does mean that operators should cooperate and bring about a national infrastructure to provide services, broadcast signal, etc, to the local networks. PTT Telecom has already offered such a facility, but competition would be served if cable operators remain independent. It is unclear whether cooperation with the electricity companies will lead to such an infrastructure. Will the electricity companies' networks be broad enough to transport all video signals? The minister has pointed out that investments can be made, but within the geographical boundaries of the permit of the party involved and without cross-subsidies.

This might mean that cable operators are not allowed to create a national infrastructure beside the electricity companies' network.

5. Conclusion

The cable-television networks in the Netherlands have evolved well into high-quality distribution networks, profiting from the support of local government. It is necessary that they evolve further into networks providing different interfaces and value-added services. So far, a few links have been missing to bring that evolution about:

- an active policy supporting the CATV operators as competitors to PTT Telecom;
- an active policy to support experimentation and cooperation between CATV operators and service suppliers, and to set flexible standards for terminals offering access to new value-added services, such as closed access, selected access and information transport.

The national government seems reluctant to take on this role. Hopefully the VECAI or a consortium of some larger cable operators, such as ADEM, will be strong enough to sponsor the necessary developments.

Notes and references

[1] This means that approximately 500,000 households do not have a subscription. Of those 500,000 about 100,000 have a satellite dish.

[2] H. Schaffers, S. Maltha and G. Fahrenkrog, De Economische Betekenis van Telecommunicatie voor Nederland (The Economic Value of Telecommunications for the Netherlands). STB-TNO, Apeldoorn, April 1993.

[3] The smallest operator (CABO, in the village of Bovensmilde) operates one network with 203 subscribers, the largest operator, Casema, operates networks in 45 municipalities with a total of 675,000 subscribers (VECAI 1991). These figures will have changed because of the concentration of electricity companies and because of take-overs.

[4] A new regulation of electricity companies has led to a concentration into large regional companies. As a consequence, electricity companies such as the EGD in the north of the Netherlands became large cable operators owning or controlling different cable systems. Another large cable operator is Casema. A large share-holder of Casema is KPN, the Royal Dutch PTT, of which PTT Telecom is an important subdivision. KPN recently announced that it would increase its shares in Casema.

[5] Talks between the electricity companies and a few CATV operators, resulting in a provisional consortium in Telened, had started even before the minister made her statement.

[6] Annual Cable Conference 1993.

[7] Recently Filmnet started a second pay-TV channel.

Telecommunication: Exploring Competition
H. Williams and M. Borman (Eds.)
IOS Press, 1994

Chapter 3

ISDN as a network innovation: experiences of the ISDN Partnership Programme

Frank Stöckler
Wissenschaftliches Institut für Kommunikationsdienste, Bad Honnef, Germany

Abstract: This paper is based on the results of a two-year evaluation of a partnership programme that was designed to promote the use of integrated services digital networks. It focuses mainly on the diffusion process and market strategy, and analyses the effects on organisational structures and communications behaviour. New marketing tools appear to be necessary to foster the diffusion of this complex network innovation. The programme succeeded in reaching users by combining intensive market communication with the promotion of practice-relevant applications and a search for powerful facilitators. Implementation will bring about changes in working conditions, the nature of employment, the skills profiles of employees and communications behaviour.

1. Introduction

With the digitalisation of central exchanges and the inclusion of integrated services digital networks (ISDN) in all European countries, a complex information communications technology (ICT) innovation has been introduced. Services and applications based on ISDN offer solutions for communications as well as new applications and new products. ISDN creates connections between the fields of telecommunications and data-processing applications that hitherto were separate. Examination of the innovation process of narrowband ISDN is of particular importance because it offers preliminary insights on the diffusion and use of integrated digital networks.

The aim of the paper is two-fold. First, it will analyse the role of a complex ICT innovation like ISDN and the success of the promotion of ISDN applications as an instrument of a broader ISDN-marketing strategy. This analysis includes the decision-making process for the implementation of ISDN and experiences of users and providers in this practical field. Second, it will deal with changes of organisational and communicational structures in these projects. It seems that depending on the way in which new digital ICTs are implemented, innovations can occur at all levels of an organisation.

It has been shown that ISDN offers opportunities for organisational rationalisation in a traditional manner by increasing work efficiency and by saving costs. However, ISDN is also a source of new forms of rationalisation. Separate office and administration processes for transmitting and communicating information and for decision-making processes, which previously were split up and handled with a time delay, can now be combined. In addition, ISDN-based ICTs can facilitate cooperative work in enterprises. This type of 'communicative' rationalisation not only reduces waiting periods and coordination problems but also changes workplace layout, working processes and thus the educational and training standards required of employees.

The management of innovations is primarily a process of social interaction. For effective innovation in terms of the diffusion of new telecommunications services such as ISDN, it seems increasingly important to initiate a dialogue between vendors, enterprises and users. Efficiency and process-orientated thinking will be increased and organisations will become better qualified to cope with an ever changing external environment in an active and flexible way.

2. ISDN as a basic innovation for new ICT services and products

2.1. ISDN: product innovation versus process innovation

In the description of technical innovations, a distinction is often made between product innovation and process innovation. While product innovation is directed towards the innovation of goods for the customer, process innovation is oriented to the innovation of production and organisation processes [1]. Product innovation manifests itself in finished products, which sell well. You can buy and use them without any problems (plug and play). Process innovation is oriented to an enterprise or a business, and prepares the way for changes of internal procedures. Process innovation has to be developed and implemented according to the specific needs of an organisation. In the ISDN-orientated diffusion-process we have both: ISDN product innovations and ISDN process innovations. New products like ISDN PC-cards, terminal-adapters, ISDN phones and ISDN videophones were developed and brought to the market along with user-oriented solutions to communications.

When describing the potential of ISDN, the following aspects become clear: a striking feature of this technological innovation is its complexity. The basic ISDN configuration of two transmission channels (2 x 64 kbit/s) and a control channel (D-channel with 16 kbit/s) or a primary-rate access with 30 B-channels [2] offers different application solutions in the fields of:

- computer linking
- communications
- information management

2.1.1. Computer linking

Due to the high signalling speed of 64 kbit/s, which can be increased further by bundling together several B channels, access to external computers is possible. Depending on the geographical spread of ISDN networks, ISDN services will be able to be used for the processing of data records all over the country [3].

2.1.2. Communications

In the field of 'communications', ISDN telephony offers new features including, for example, a quicker call set-up, call-number identification, automatic callback, itemised billing and an improved voice quality. The simultaneous use of two B-channels enables the use of computer conferencing or video telephony.

2.1.3. Information management

The transmission and processing of all kinds of information (documents, archives and research) can be optimised by ISDN-connected workplaces and ISDN-based information-management systems are already in use. Thus, associations can use ISDN to provide information and services in the same way as, for example, travel agencies and service departments in the specialised electrical trade. ISDN will soon offer the possibility of interrogating an organisation's information, statistics, travel plans and repair instructions and of reprocessing them. ISDN applications can be used in all branches. Existing applications are:

- ISDN videophones for experts
- branch-store information systems
- sales control via ISDN
- ordering and billing systems
- ISDN-based medical communications
- remote maintenance
- services for tax counsellors

This short description shows that ISDN is a complex technical innovation and its complexity is emphasised by the fact that it can be both a product innovation and a process innovation. This complexity can be seen to hamper the diffusion of ISDN, but is perhaps offset by the potential benefits from the innovative activity which ISDN can induce. An ICT innovation like ISDN can neither be made for the shelf nor sold from the shelf. These characteristics of ISDN as both product and process innovation are important issues influencing the development of the market, for there are considerable differences between the diffusion processes of both 'innovation forms' [4].

In the initial phase of market development, the acceptance of ISDN was hindered by its product-oriented marketing. That ISDN in most cases was a process innovation needing explanation and that its implementation and use implied reorganisation processes was neglected. To reach innovators and initiate the diffusion process of a complex process innovation like ISDN, interaction between vendors, systems developers and users seems to be critical.

2.2. *The ISDN Partnership Programme – a new instrument of marketing*

Man kann sich nicht wünschen, was man nicht kennt (Voltaire) [5] (You cannot wish to have something, you do not know).

This quotation points to an essential aim of the ISDN-application promotion started by Deutsche Telekom in 1989 and a critique of previous attempts to match ISDN. Funded by 7 million DM, more than 70 cooperation agreements have been made between users, producers and Deutsche Telekom. To introduce ISDN, this 'promotion model' has proved to be necessary and reasonable and 30 partnership projects have been completed.

A special emphasis of the Partnership Programme is the testing of user-oriented ISDN solutions. Each partnership scheme must meet three criteria: they have to be innovative, economic and have an obvious multiplier (or facilitator) effect.

Following an intensive exchange of experiences with customers, the aim was an improved customer-orientation and market-orientation, as well as a continuous improvement of product and service quality. In 1991, Deutsche Telekom commissioned Wissenschaftliches Institut für Kommunikationsdienste (WIK) to evaluate selected application projects. This evaluation has informed the analysis in this paper.

3. Evaluation of promoted ISDN applications

Since mid-1991, WIK has been concerned with the evaluation of some ISDN-application projects. The Institute, which was founded by the Deutsche Bundespost in 1981, became an independent organisation in 1991. At present, WIK is the biggest research institution for economic and social issues in the field of post and telecommunications in Germany. A new research field of the Institute is technology assessment which includes the evaluation of ISDN-promoted projects.

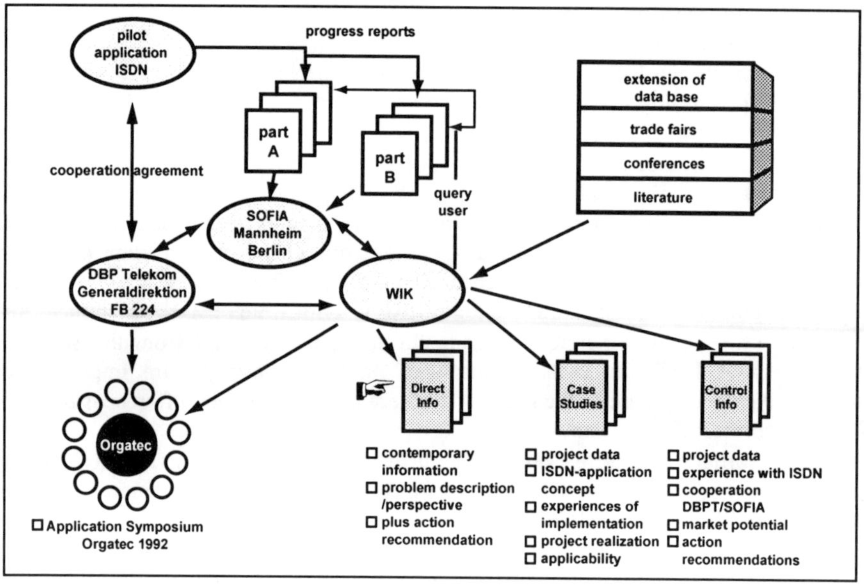

Figure 1: Evaluation process of ISDN projects

The implementation of innovative communications technology like ISDN implies questions beyond the technical dimension, because the introduction of technical systems in organisations has to be considered as part of an organisational and social process. For this reason, the initial stages of evaluation are not focused solely on the technology but include observation of the development and effect of organisational change.

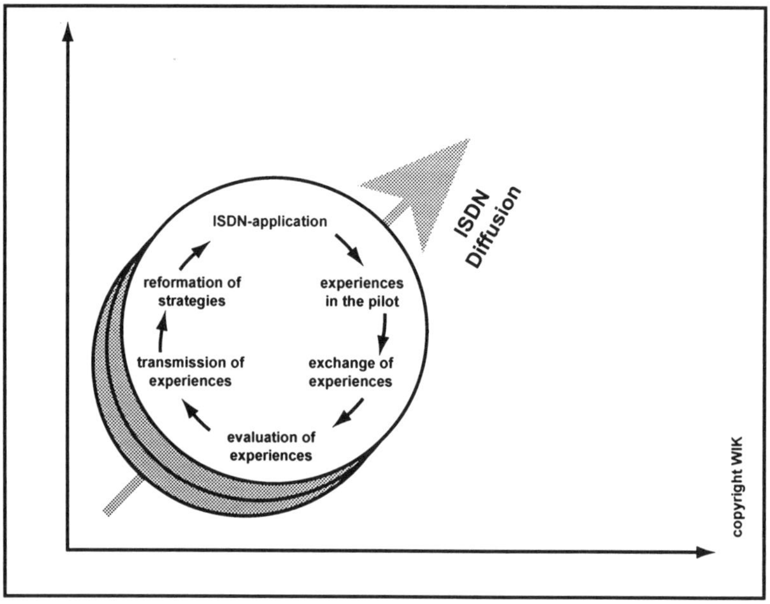

Figure 2: Exchange of experiences for the diffusion of ISDN

Data on the ISDN configuration, quality of products, organisational changes and experiences of the cooperation and service provided by Deutsche Telekom were gathered and analysed by a two-stage survey, expanded by telephone interviews and personal discussions. The survey dimensions were:

- ISDN-application concept (hardware and software configuration)
- implementation of ISDN application
- organisational changes
- ISDN in continuous operation
- perspectives of ISDN applications
- cooperation with Deutsche Telekom

Publications and presentations at conferences have provided constant information on experiences with ISDN in business-use during the run of the project. Particular emphasis was on the diffusion of ISDN and drawing up an experience and diffusion circle. This circle develops the promotion concept by evaluating and passing on user-oriented knowledge. New ISDN applications as well as applications that demonstrate forward-looking development lines for potential users are distributed, thereby encouraging the diffusion of ISDN.

The ISDN innovation process and the innovative use of ISDN is seen as a recursive process [6] which is characterised by numerous feedback loops, iterations and intersections in the innovation phases.

The aims of the partnership programme were two-fold: the implementation of ISDN-based innovations through the promotion of pilot projects, and the development of a 'learning by doing' procedure. The complexity of ISDN as a process innovation seems not to concur with the linear 'invention-innovation-diffusion-model', and thus the 'innofusion model' [7] seems to be more relevant. In the idealised course of adoption, ISDN-application promotion intervenes in such a way as to support the innovators financially and organisationally.

4. Experiences of the evaluation project

The next section deals with the participants in the invention phase and with the assessment of ISDN before its implementation in enterprises. An assessment is given of 'promoted' users on their function as facilitators for the diffusion of ISDN. The following section looks at existing and possible patterns for the rationalisation of office work. The thesis is that the use of ISDN-based ICTs is accompanied by a kind of 'communicative rationalisation'. It has to be pointed out that due to the limited observations of promoted ISDN projects, only initial reflections are given and these need further critical examination.

4.1. 2B or not 2B?

Table 1.

For ISDN	Against ISDN
• cost reduction	• initial innovator risk
• high transmission speed	• information deficit
• improvement of quality	• initially, lack of coverage
• improvement of service	• hardware and software problems
• improvement in communications	• problems with implementation
• rationalisation and efficiency of organisation	• organisational problems
	• technical problems
	• lack of experts on the spot

In most cases, the idea of ISDN implementation came from the users themselves ('users as innovators'). Information was obtained at specialist fairs or by direct contact with Telekom where similar telecommunications applications already exist. The main arguments for or against the implementation of ISDN are emphasised by Table 1.

The main arguments in favour of implementation were the expected reduction of communications costs and the possibility of quicker data transmission with the consequent effects of a rationalised and improved service. Arguments against the introduction of ISDN were, above all, the anticipated innovator risk, information deficits and the lack of coverage, particularly in the initial projects. In some cases in particular, the long waiting periods for ISDN access and the uncertainty over the development of tariffs at Telekom were listed as counter-arguments. In most cases, the decision by Telekom to give financial, technical and organisational support – particularly in the initial phase of ISDN – was decisive for the decision to adopt.

Nearly half of the projects evaluated were initiated by third-parties, for example, software and consulting enterprises, producers and universities. Consequently, three sorts of facilitators for ISDN diffusion can be defined (see Figure 3).

In this context, opinion leaders are those persons, enterprises or organisations who were accepted by potential users as 'innovators'. Their innovative behaviour is seen as an example and can initiate an imitation process across a peer group, whether within the same company or across different firms in the same sector.

Cooperation with the so-called 'experts', mainly specialists who inform the users about the idea and do the required adaptation work, seems to be important for the acceptance of ISDN. An incentive for these facilitators is privileged access to the projects as well as the collection of practical and organisational knowledge for their own product and services developments.

The third group of facilitators are those organisations that have the power to diffuse ISDN-based innovations across an inter-related and inter-dependent network of firms. Like a 'spider' in the net, this sort of facilitator holds all the reins in the communications network. The user has to switch to the new technology to make use of the advantages of special services, for example, data-processing services for tax counsellors.

There is no question that the facilitators are an important target group for ISDN marketing. The contact and support of this kind of facilitator could be very useful for the further diffusion of ISDN-based ICTs.

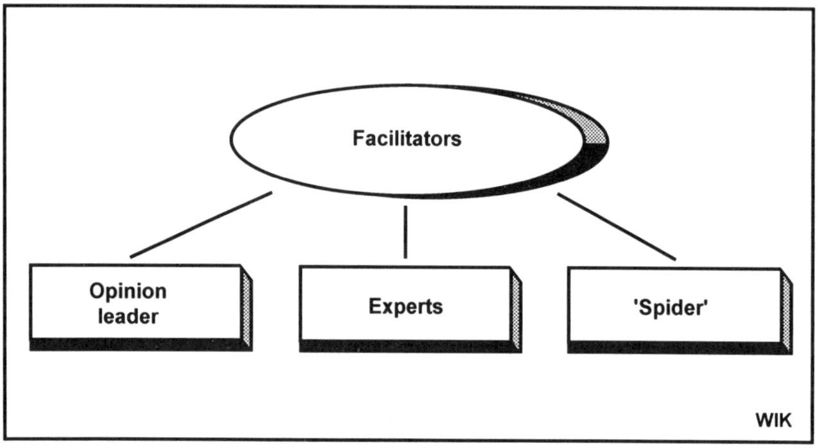

Figure 3: Facilitators of ISDN applications

4.2. *Bye, Bye Taylorism?*

An important factor for companies is the rationalisation of working procedures. The importance of working conditions and procedures is well-known from the description of the division of labour by Emile Durkheim on the macro-level of societies, and from Taylor who elaborated this phenomenon on the micro-level of companies.

The effects of ICTs on the individual, organisation and society are based on the common and extensive significance of 'information and communications' in business and everyday life. The implementation of 'ISDN-based ICTs' and the associated integrated services for the networking of enterprises and information management will not be without effects on business and communication structures. Communication and organisation structures are affected differently by the new features and services of ISDN. Previously separate office-work tasks, for example, those of a typist, telephonist, office messenger, were integrated by modern telecommunications (personal computer, telephone, e-mail). The mode of rationalisation and action varies, depending on the service used, plant size, structure and density of internal and external communications networks, as well as on the degree of organisational networking [8].

Preliminary evidence from the evaluation project demonstrates some of the changes that can occur in office communications and lead to changes in working procedures. First, is the example of cooperation with the Badischer Sparkassen- und Giroverband, a regional association of savings banks. The association acts as a consulting, information and communications head office for approximately 60 savings banks. With the networking of PC-workplaces in the head office and external access via a combination of ISDN-based telecommunications equipment and the main computer, internal as well as external information transfer could be initiated more quickly and communication became both more intensive and extensive.One particular cost advantage can be attributed to information that is forwarded by e-mail. Copy and mail costs are saved. In the workplace, a kind of qualified case-processing was introduced: with local networking, employees have extended access to the information stock. The handling of intra-organisational and inter-organisational transactions is expanded by integrated speech, picture and datacommunications. Important effects of such classical rationalisation are:

- improvement of direct-mail addressing
- avoidance of down time
- reduction of processing time
- fewer letters
- time-saving on investigations

A second example is that of the 'Tax Counsellor Project'. In cooperation with DATEV [9], a nationwide service-provider for tax counsellors, the advantages of ISDN were used to integrate branches more effectively. For example, the association-owned service centre in Nuremberg has switched ISDN access for tax consultants to the mainframe computer. Offices equipped with ISDN PC-cards, can handle enquiries and calculations quickly and cost-effectively. Data transfer no longer has to take place at night as it did previously. The processing of tax files is no longer split up, and is now more cost-effective. For employees in these projects, the work has become more demanding and more responsible and requires assistance. From these examples, the thesis that technical and organisational innovations require other and often higher qualifications, and do not result in less-qualified service work, is confirmed.

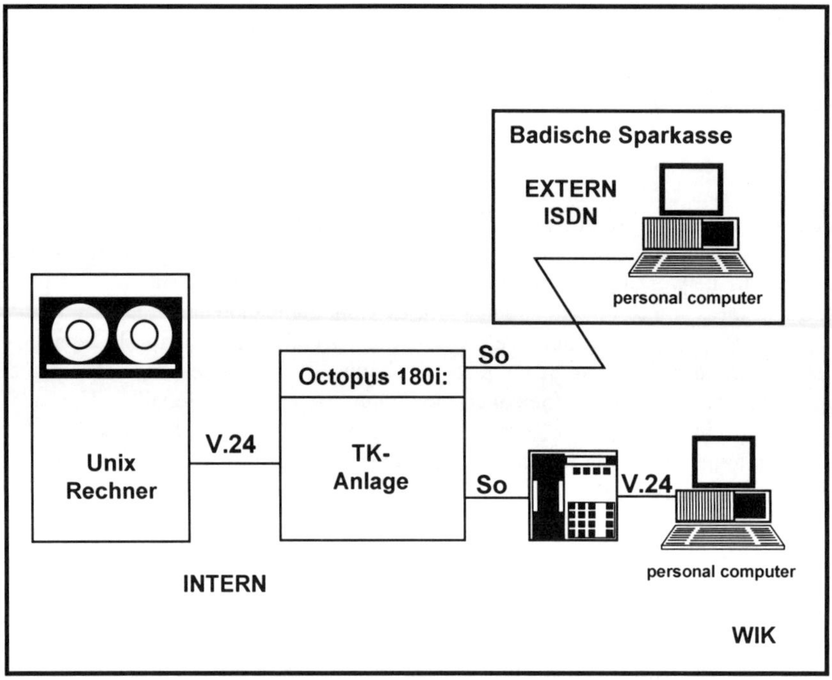

Figure 4: Exchange of association information via ISDN

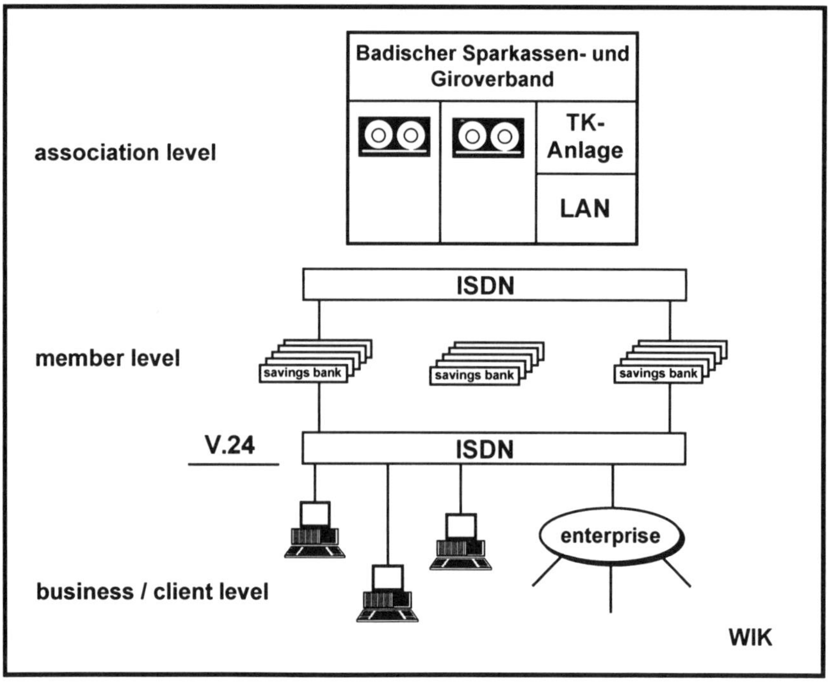

Figure 5: Communications solution for the association

In a third project, the improved availability of data and information supported a new form of group-work. With the use of ISDN ICTs in the workplace, several work processes could be integrated (process-integration) and, as a result, the method of working changed. Activities like the buying and selling of goods and their administration for production that previously were separate, are now managed by a team connected via ISDN. This method of working is also occurring in the development of new products. There are plans to introduce this new form of cooperative work to the whole organisation. In other words, the previous regulated and hierarchical organisational structure is changing to a 'communicative organisation'. This new form of rationalisation improves the relationship between management and administration. It could be argued that the scope for action and for decision-making by employees also changes [10].

The problems and effects of Tayloristic rationalisation are well known. After the Second World War, the Japanese Taijchi Ohno [11] noticed that the knowledge of American assembly-line workers could be used to improve the production process. Well-known as 'Toyotism', he introduced production-teams and group-work at Toyota and, as a result of this reorganisation, product and service quality were improved.

With the implementation of ISDN-based ICT, developments similar to those on the factory floor can be seen in office organisation. It seems that 'communicative' rationalisation is able to eliminate the frictional losses of Tayloristic rationalisation. Contrary to Taylorism, there is less differentiation between those who 'think' and those who 'work'. The strengthening of cooperative working conditions emphasises the responsibility and creativity of the employees. The effects are: firstly, that relatively high production costs and loss of production can be prevented and, therefore, costs of quality management reduce; and secondly, that new tasks can be managed in an integrated and flexible way.

The limitations are that in this phase of the ISDN-innovation process, relatively little is known about the particular influences of ISDN-based ICTs on the rationalisation of working conditions. Initial observations are based on the adoption of individual applications in an individual unit (micro-level innovation) [12]. According to this, ISDN-based ICTs will effect communication and organisational structures in many ways. Further interdisciplinary research is needed in this field.

5. Conclusion

For a complex multi-dimensional innovation like ISDN, new ways of market development and sales promotion have to be developed. With the promotion of ISDN applications by Deutsche Telekom, innovators for practical applications were found. With organisational and financial support, existing barriers to the adoption of ISDN innovations were overcome. With intensive customer orientation, not only promoted applications but also the whole service and ISDN products underwent continuous improvement. Cooperation between hardware and software producers, Telekom and users proved to be positive, not only for the development of innovative applications (users as innovators) but also for the market orientation of Telekom.

However, it becomes obvious that beside traditional patterns of rationalisation, ISDN-based ICTs influence working and communications processes in their work environment. It seems that the implementation of ICTs based on ISDN leads to reorganisation on all levels within the organisation. For the efficient use of ISDN-based ICTs, it seems increasingly important to analyse the existing process of rationalisation and develop new 'communicative' forms of rationalisation. Product and service quality will be increased, and the enterprise will become better qualified to cope with an ever changing and turbulent external environment.

The projects that were examined did not provide any findings about how far the actual nature of employment and the skills profiles of employees is being threatened by the rationalisation potential of the new ICTs. The expected increase in networking within and between enterprises – even into private households – raises questions about data security and data-privacy protection, which are going to become important with the increasing number of ISDN users. This is an issue that warrants further and detailed investigation.

It has to be pointed out that, due to the limited observation of promoted ISDN projects, only initial impressions have been presented. These need further empirical examination. This means that further research is necessary for a precise estimate of the whole spectrum of effects and for assessment of the change in organisational processes following new ISDN-based ICTs.

Notes and references

[1] As product (or supply) innovations, the introduction of new or improved products or services are described. Procedure (process) innovations are changes in technological working processes. They concern the improvement or total innovation of procedures. N. Thom, Grundlagen des betrieblichen Innovationsmanagements. (Basics of innovation management in companies.) Hahnstein Verlag, Königstein im Taunus, 1980, pp 32–37.

[2] 30 x 64 kbit/s plus 1 x 16 kbit/s.

[3] For example, those processed by tax counsellors.

[4] Eidgenössische Technische Hochschule Zürich (Ed.), ETH Research Project 'MANTO', Working Group 'Diffusion', working paper 3.11. Diffusion of new applications in telecommunications, Zürich, 1986, p 7.

[5] M. Zeller, Customer comes first. Paper presented at Telematica Conference, 19–21 September 1990. Stuttgarter Messe- und Kongreß-Gesellschaft mbH (Ed.), Stuttgart, 1990, pp 83–92.

[6] J. Asdonk, U. Bredeweg and U. Kowol, Innovation als rekursiver Prozeß. Zur Theorie und Emperie der Technikgenese am Beispiel der Produktionstechnik (Innovation as an ongoing process. Theorectical and empirical aspects of technical development in production technology), *Zeitschrift für Soziologie*, Vol. 20, No. 4, August 1991, pp 290–304.

[7] J. Fleck, Innofusion or Diffusation? The nature of technological development in robotics. PICT Working Paper Series No. 4, Edinburgh, 1988, p 1.

[8] see D. Garbe, Kostensenkende Datenautobahn - ISDN senkt die Kosten (Cost reduction by ISDN data highways), *Business Computing*, November 1992, pp 88–92.

[9] Datenverarbeitungsgesellschaft der Steuerberatenden Berufe (data processing association for tax counsellors).

[10] K. Monse and H.-J. Bruns, ISDN-Einsatz in einem Handelsunternehmen- Einige Anmerkungen zur Gestaltung der Organisations- und Kommunikationsstrukturen. (Implementation of ISDN in a trade company: aspects of the changing organisational and communication structures.) Paper presented at the Online Conference, Congreß II, Symposium II–4, Hamburg, 1992 pp II.29.01–II.29.21.

[11] Taijchi Ohno, Das Toyota-Produktionsprinzip. (Principles of the Toyota-production System.) Campus Verlag, Frankfurt, 1993.

[12] Juhani Iivari, From a Macro Innovation Theory of IS Diffusion to a Micro Innovation Theory of IS Adoption: an application to case adoption. In: D. Avision, J. E. Kendall and J. I. DeGross, (Eds). Human, Organization and Social Dimensions of Information Systems Development. IFIP Transactions A: Computer Science and Technology, North Holland, Amsterdam, 1993, p 296.

Telecommunication: Exploring Competition
H. Williams and M. Borman (Eds.)
IOS Press, 1994

Chapter 4

ISDN: a technology of the past for the telecommunications of tomorrow?

Gerhard Fuchs
Akademie für Technikfolgenabschätzung in Baden-Württemberg, Stuttgart, Germany

Abstract: In the late 1980s, the integrated services digital network was being called a technology of the past – even before it was introduced as a widely-available commercial service. However, it has become an unexpected success story. This paper analyses some of the reasons often quoted for its misguided implementation. Using the comparative method developed by Charles Ragin, it is argued that most interpretations are not very convincing. It goes on to assess the present state and future of the integrated services digital network in Europe. It is shown that the European Commission will play a decisive role in enabling its eventual success.

1. Introduction

The integrated services digital network (ISDN) was called a technology of the past even before it had become widely available as a commercial service. It seemed not only to be a solution in search of a problem (Newstead) [1], but also a service looking for customers ('Innovations Subscribers Don't Need'). ISDN might have been the last universal project of the old monopoly network operators, but it also constituted the first serious network project the Commission of the European Communities (CEC).

The initial hopes and expectations linked to ISDN without any doubt have largely failed. Just a few years ago ISDN was identified as an important ingredient of the so-called 'age of freedom' (Sussman) [2] and was considered by both national network operators and the CEC as the decisive step towards the information age. In both the national as well as the European arena, ISDN has now been discarded. Within the CEC, the ISDN programme is considered by many as an ill-fated policy, and talking to national actors, nobody feels responsible for the far-reaching ISDN programmes. But on the other hand, the EU discusses broadband infrastructure plans that resemble narrowband ISDN plans and, on the national level, there are still network operators and equipment producers who consider the future of network development in terms similar to those discussed in the original ISDN schemes – only skipping N-ISDN as an important and necessary intermediate step (see Figure 1) [3].

What can be seen in the case of ISDN seems to be that evaluations are jumping from one extreme to another. After over-emphasising the importance of ISDN for so long, it has now become in vogue to generally downgrade ISDN. Even the professional telecommunications-policy community seems to have lost any significant interest in ISDN, in spite of the fact that the real implementation of ISDN is only about to start. It is obvious that ISDN will be less than was originally expected, but it may be more important – especially for public network operators – than many concede at the moment.

To get a better understanding of the impact of ISDN, this paper explores what have been the dominant interests in developing ISDN. On the basis of this discussion, it will be demonstrated that not only was ISDN developed with diverging interests in mind, but that this has also led to different implementation strategies. The discussion of ISDN has suffered from being directed towards a more or less undifferentiated concept of ISDN. Furthermore, ISDN is often discarded on the basis of arguments about its origins. However, problems related to origins *might* be irrelevant to the context of implementation.

An evaluation of ISDN has to take into consideration the variety of strategic aims and issues associated with ISDN.

1.1. Just a reminder about what ISDN meant in the beginning

The original ISDN plans centered around a couple of general technological, economic and strategic issues, that can be summarised with the keywords of integration and digitalisation. *Integration*: telecommunications networks of the pre-ISDN era were constructed to transmit information. They were constructed to transmit information in a specific form, for example, the telephone network to enable acoustic communication between individuals, the telex network for written information exchange. The telecommunications networks were not planned for general transmission of information. Thus, they were originally service-specific networks: each new service got its own new network. This has changed and increasingly new telecommunications services are offered in existing networks. Telefax, videotext and certain datacommunications services are using the telephone network; teletext and specific datacommunications services are offered via the traditional telex network, that has been transformed into an integrated data network. The telecommunications networks have thus become service-integrated networks.

Digitalisation: the telephone network is still the world's most extensive telecommunications network. It used to be an analog network. The digitalisation of the analog networks makes it possible to integrate digital-data services. Once the telephone network is digital, it can provide the broadest basis for the integration of all voice and non-voice services in a single network.

Thus, it becomes possible to create an uniform 'telecommunications highway'. This highway can be viewed as an infrastructure that serves the transmission of a variety of telecommunications services like telephone, telefax, teletext, videotext, data dialogue, batch traffic, picture communication, and so on.

Some typical ISDN features were supposed to be: one subscriber number, automatic terminal access, two connections or the use of two different services at the same time, shorter connection set-up time, change between services during a call, portable, 'plug-in' terminals, change between devices without changing services, semi-permanent connection, closed user-groups, call-progress signals, blocking of outgoing calls for all services, itemised recording and printout of call duration and charge, status inquiry, continuous activation. ISDN terminals were supposed to be multifunctional and to replace as many as four or five traditional pieces of equipment (like fax machines, teletext machines, typewriters, computer terminals). Finally, the advantages of ISDN have been characterised as a higher transmission speed for data (compared with existing telephone networks), a more economical use of the various telecommunications services (drastic reduction in user costs), a more comfortable use of the various services, universality of the access for text, voice, data and image communication (Griffiths) [4].

In this sense, ISDN can be categorised as a mixture of a process innovation – the digitalisation of the network, and a product innovation – existing services could be altered and new services offered. Digitalisation does not make it necessary to integrate services. It is, however, to a large extent the precondition for creating new telecommunications services.

The CCITT [5] I–Series Recommendations defined ISDN as 'a network, in general evolving from a telephone IDN (integrated digital network), that provides end-to-end digital connectivity to support a wide range of services, including voice and non-voice services, to which users have access by a limited set of standard multipurpose user-network interfaces.' [6]

A lot of these characterisations now look quite outdated. Several of the characteristics mentioned above are offered outside the ISDN framework. But nevertheless, ISDN has an overall importance that goes beyond the outdated initial programmatic statements. Technically, ISDN meant giving up the idea of separate networks for separate services and the complete digitalisation of the telecommunications networks. Strategically, ISDN meant a strengthening of a strong, national telecommunications industry in order to protect it against assault by the computer industry.

The CEC had the same reasoning for its ISDN-support strategy: besides the strengthening of its own role in telecommunications, it intended to support European industry and develop a homogenous, integrated European telecommunications infrastructure. These were bold plans on both the national and international level. Integration and digitalisation were used as key concepts with which virtually every conceivable need of the telecommunications user could be satisfied. Unlike other national initiatives, the discussion about ISDN and its further development into a B-ISDN represented an unprecedented long-term planning and investment effort, that was not only proceeding in the national arenas, but was also oriented to the global development of networks. In this sense it constituted a first and last planning exercise of the international community of monopoly network operators. In a similar manner, it constituted the first and most probably also the last broadband telecommunications-network initiative by the CEC [7].

2. The introduction of ISDN

There has been some mystery about what actually drove the introduction of ISDN. Besides the not very illuminating but nevertheless generally shared sentiment, that ISDN was a technology-driven process and therefore failed, because it could not work in a free-market environment [8], three interpretations seem to prevail in the social-science literature [9]. The first interpretation looked upon ISDN as a plot by the old network monopolies to secure their status and fight off competition from the computer industry. A second dominant interpretation again saw ISDN as the result of a plot, but this time by some multinational telecommunications-equipment producers, who had a strategic long-term view about their business prospects and who also smelled a competitive threat coming from the computer industry. ISDN seemed to provide them with substantial orders over a long period of time and would also tie the network operators to them, because the computer industry would not be able to compete fiercely on the ISDN market. Finally, some interpreted the ISDN process primarily as part of a strategy motivated by an industrial policy, that wanted to use ISDN as an instrument to upgrade infrastructure and provide a competitive advantage for the home industry. All three interpretations can rally impressive arguments behind them and they should not be easily discarded [10].

Examination of what actually happened and when in the implementation of ISDN, rather than unfulfilled expectations, provides a rather complex view. In order to justify this interpretation, a systematic comparison is made of ISDN introduction, applying the comparative method developed by Ragin [11].

The three main explanations can be described and simplified. In them, three causal determinations have been established: industrial-policy motives (A), the behaviour and existence of public monopoly network operators (B), and strong national equipment producers (C). And in view of the interest in innovative and long-term commitment, nations in which commercial ISDN services (in the standardised CCITT version) were available before 1 January 1990 are also examined [12]. This will be the output variable Y.

The following is a short explanation of the variables. Since Ragin's comparative method is being applied, all variables have to be codified in the existent/non-existent mode.

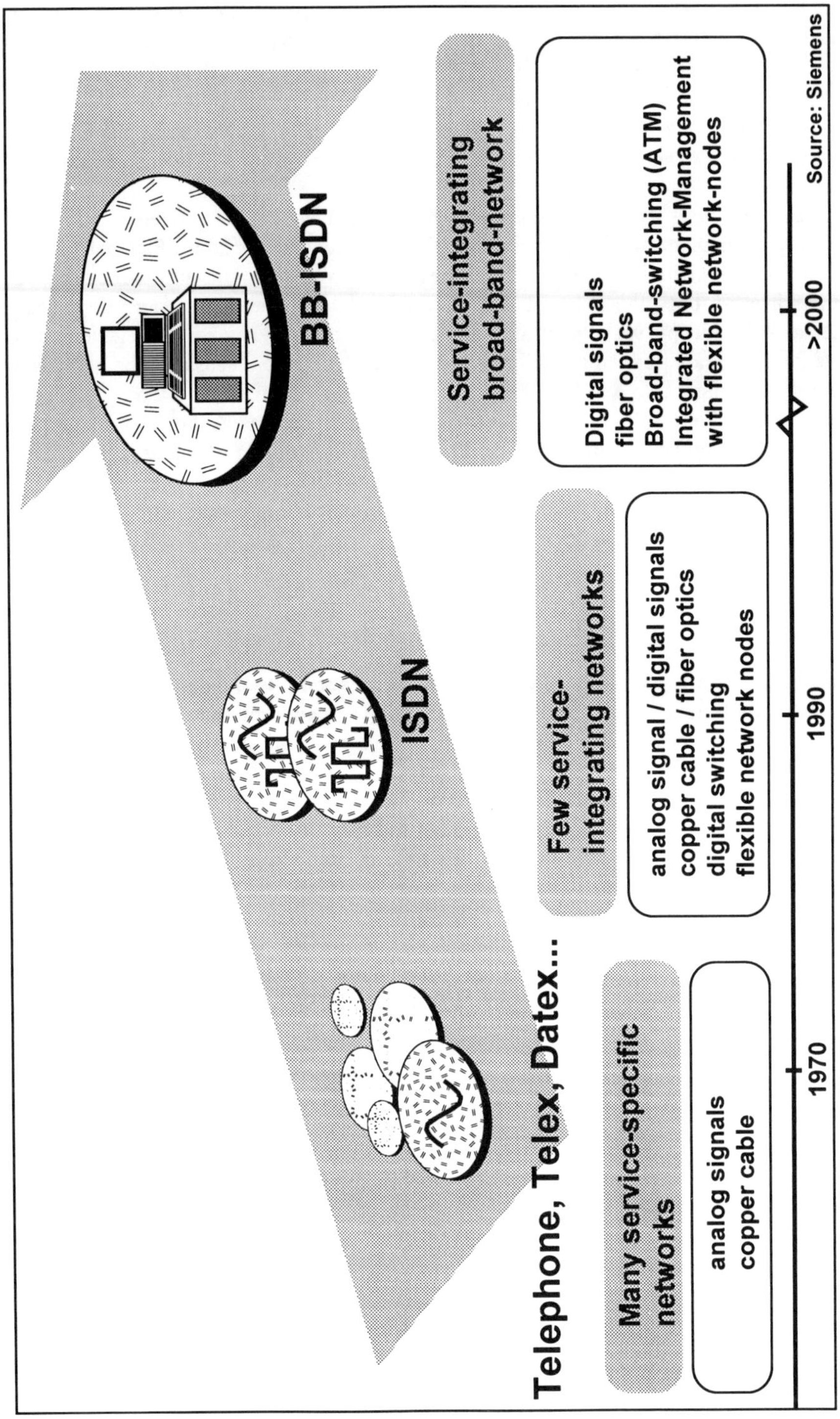

Figure 1. Network evolution

Without doubt this implies some serious simplifications, which have to be accepted for the sake of making meaningful comparisons. However, this is not a problem with respect to the existence of a monopoly network operator. For codification, the base-year 1986 is used – the year when the CEC's ISDN recommendation was issued and more reliable time tables for its introduction in a variety of countries were available. The industrial-policy motivation in telecommunications has been coded on the basis on a study by DIW [13] [14]. The problem of national equipment producers has been approached in the following manner: in the early 80's there was already a widespread consensus that only very few corporations would be able to come up with the billions of Deutsche Marks seemingly necessary to develop a new generation of ISDN-capable digitised switches. Therefore, it was estimated that only companies commanding at least 8% (this figure was reinterpreted and changed a couple of times) of the world market will be able to survive in the long run. The existence or non-existence of such a corporation in a national market was coded in this way. Table 1 shows that today, only a small number of companies have been able to sell and install ISDN systems.

Table 1. ISDN technology based on CCITT standards in operation worldwide

Public Switching Suppliers:	Technology: Systems supplied:	Countries served:
Siemens	EWSD	11
Ericsson	AXE	11
Alcatel N.V.	1240	10
Fujitsu	Fetex 150	5
Northern	DMS 100	4
Alcatel CIT	E10-B	3
AT&T	5 ESS	3
NEC	NEAX 61	2
GPT	System X	1
Nokia	DX 200	1

Status: March 1993. For further information see Table 8 in the appendix.

Table 2. Start of commercial ISDN operations [15]

Year	1988	1989	1990	1991	1992	1993*
Countries:	France USA Japan Germany	Belgium Switzerland Australia Canada Singapore		UK	Denmark Netherlands Italy Portugal Hong Kong Spain Finland Ireland	Austria Sweden Columbia Taiwan Norway

* already in operation or planned

Most problematic has been the output variable. If the various publications on ISDN are studied, conflicting data are given on when commercial ISDN was introduced – not to speak of data on ISDN penetration. The following data seem to be the most accurate (Table 2).

For the systematic comparison, all countries that introduced ISDN before 1986, plus a sample of countries that intended to introduce ISDN at a later date were examined. The preliminary results are given in Table 3.

By using some simple logical procedures of Boolean algebra on which this exercise is based, interesting conclusions can be arrived at. One is that the early success of ISDN is dependent on different combinations of factors and is not the result of a one-dimensional simple configuration.

Table 3. Truth Table: The introduction of ISDN and its determinants

Condition			Outcome	Number of cases	
A	B	C	Y		
0	0	0	0	1	(UK)
0	0	1	1	2	(Canada, USA)
0	1	0	0	2	(Denmark, Netherlands)
0	1	1	?	?	n.a.
1	0	0	0	2	(Spain, Italy)
1	0	1	1	1	(Japan)
1	1	0	1/0	5	(Belgium, Netherlands, Switzerland, Australia, Singapore)
1	1	1	1	2	(France, Germany)

* For more details on variables and cases see Table 7 in the appendix.

For the success of ISDN (I), four different combinations of conditions could be verified where: [16]

I = ABC + ABc + AbC + abC

An early introduction of ISDN can be observed in any of the four combinations and none as such is more 'true' than any other. ISDN could be introduced: in a situation in which there were or were not rival network operators; in cases in which industrial-policy motivation was present or absent; and in cases in which there is or is not a dominant multinational corporation that would have liked to use its home country as an ISDN showcase for sales abroad.

The 'failure' to introduce ISDN at an early date is both more complex as well as more straightforward in its results. ISDN 'failure' (i) can be seen in conjunction with the following configurations:

i = aBc + abc + Abc + ABc

This formula can be simplified:

i	=	aBc + abc	=	ac
i	=	abc + Abc	=	bc
i	=	ABc + Abc	=	Ac
i	=	aBc + ABc	=	Bc

i = c (a + b + A + B)

It can be concluded that c (the non-existence of a strong national equipment producer) is a necessary condition. In other words, in countries that have no important producer, there is a high likelihood that there will be no early ISDN deployment, unless it is substituted by a strong political initiative. The most interesting cases then, are those linked to the configuration ABc. In some cases there is a positive result and in others a negative one. For those countries, the specific strategic considerations of the main actors that led to the ISDN decision, will have to be more closely examined.

As a result of the discussion so far, it can be concluded that equipment producers clearly play a decisive role in understanding the whole process. All countries that feature a dominant player on the world market have also been successful in supporting an early introduction of ISDN. The lack of an important international producer can only be counteracted by a strong alliance of industrial-policy motives and a network monopoly. The industrial-policy motives in these cases may be infrastructure oriented or related to the development of the domestic production facilities of the big players. But at any rate, it becomes obvious that the introduction of ISDN is driven by a variety of motives and not by one of the shorthand explanations usually offered.

3. ISDN implementation strategies

So far, this paper has presented ISDN as something monolithic and as though each actor introducing ISDN would have the same basic idea as to what ISDN is all about. But in fact ISDN means different things to different people. One indication of this is the distribution between basic access (BA) and primary-rate access (PRA) in the different countries [17]. At the moment ISDN is being introduced using different strategies: some operators have high introduction profiles; others cleverly hide their ISDN offers. Furthermore, the variety of 'successful' countries might also speak in favour of very specific approaches. Some network operators treat ISDN as one service among others (and, therefore, put less emphasis on it), while others treat it as a national infrastructure concept (mainly the political actors in the game.) Finally, a classification of strategies might further qualify the result of the systematic comparison and dispel the impression that pure contingency may be at work here.

With respect to the distribution of BA:PRA, different groups can be distinguished right away: a group in which there is no or very little PRA at all; a group where there is in relation to BA a considerable degree of PRA (like Australia, Denmark, UK, Italy, Netherlands); and a group in which there is a balanced relationship between the two (10:1 up to 25:1) More broadly, a distinction can be made between four different strategies.

1 *ISDN islands*. Based on perceived need and interest by the network operators, ISDN islands are being created. A good example is the McDonald's ISDN in the USA, which connects branches of this franchising company. This strategy operates with some proprietary norms and standards and it is usually difficult to connect ISDN islands with one another, especially if the operators use equipment/switches from different suppliers. Thus, closed user-groups come into existence which communicate via ISDN among themselves, but are

separated from the general public network. This concept is not solely market-driven, because some Bell operating companies (BOCs) in the US clearly see other network concepts as being more beneficial to them. Also, in these cases the regulatory framework of American telecommunications prevents the operators from making full use of the potential of ISDN (see Table 4). In market-driven cases it is also to be expected that compared with BA, the emphasis on PRA is striking. PRA is only oriented to the needs of the (larger) business users.

Table 4. Number of ISDN accesses (PRA/BA) as at March 1993

Countries	Suppliers	Basic Access	Primary-Rate Access
Australia	Ericsson	7000	1000
Austria	Northern, Siemens	192	
Belgium	Alcatel, Siemens	1000	20
Canada	Northern, GTE	750	20
Columbia	Siemens	5032	62
Denmark	Siemens, Ericsson, DIAX	140	62
Germany	Alcatel, Siemens, DeTeWe Telenorma	122929	8331
Ireland	Alcatel, Ericsson	34	
Italy	Alcatel, Ericsson, Italtel	7000	130
Japan	NTT	100000	2500
Korea	ETRI	1300	
Malaysia	Fujitsu, NEC, Ericsson	n.a.	n.a.
Mexico	Ericsson, Alcatel	200	
New Zealand	NEC	100	
Netherlands	Alcatel, AT&T, Ericsson	900	100
Norway	Alcatel	503	80
Portugal	Siemens, Alcatel	1806	14
Sweden	Ericsson	800	20
Switzerland	Siemens, Ascom, Alcatel	2700	150
Singapore	Fujitsu	750	
Spain	Alcatel, Ericsson	1900	152
Taiwan	Alcatel, AT&T, Siemens	828	
UK/BT	GPT, Ericsson	11500	6000
UK/Mercury	Northern	–	4000
USA	AT&T, Siemens, Northern	244989	2309

Table 5. ISDN access (in channels) (% of inhabitants)

Country	% of inhabitants
USA	2.2
West Germany	6.0
France	6.0
UK	5.6
Japan	2.1
Australia	6.1
Columbia	0.7
Italy	0.3

Source: Siemens

2 *ISDN-infrastructure upgrade*. The data in Table 4 provide an the interesting result – the hopes and plans initially associated with ISDN only seem to be shared by newly industrialising or developing countries. Singapore and other countries use ISDN for rapidly upgrading their general infrastructure. They also hope to gain competitive advantages if they can provide a reliable network for the domestic market and especially the foreign business community. In these cases therefore, the main interest is with the development of BA and not so much with PRA.

3 *Demand-oriented universal ISDN*. Operators like British Telecommunications still promise that ISDN will be universally available for private as well as business customers. Their strategy, however, is clearly profit and market oriented. The construction of the network follows a respective logic. This might also explain why, in countries like the UK, the proportion of BA to PRA again is lower than in other countries. PRA is clearly of use to private households and small firms.

4 *Universal-monopoly ISDN*. Countries like France and Germany, who still feature a quasi-monopolist terrestrial infrastructure, share the idea of a general upgrading of their telephone network to the ISDN standard. The spread of the infrastructure is to a large extent independent of demand, but rather follows general planning decisions. ISDN is considered to be an intermediate step towards the development of broadband networks. ISDN, however, has lost its exclusive role and has to compete with a variety of other network options. This orientation also becomes visible in the marketing strategies of the network operators. By marketing the so-called 'killer application' file transfer, France Télécom wants to persuade the private user of the value to them of ISDN.

3.1. The Commission of the European Communities as a new actor in telecommunications and the ISDN telecommunications-network initiative

The nation-oriented comparison under-rates the importance of international developments for the realisation of ISDN. One of these international developments concerns the Commission of the European Communities (CEC). The CEC had an important catalytic function in the process of transforming European telecommunications. Its activities were first of all oriented towards network planning. A new level for coordinating national-network planning was supposed to be created by the Commission. Later on the CEC concentrated on the creation of a new regulatory framework for telecommunications. The Green Paper of 1987 symbolised attempts to force the member states to implement regulatory reforms. More recently, the Commission has tried – with the use of the law of competition – to challenge national regulations. At the same time, the EU is trying to create new European institutions and structures, which might work as a stable foundation for a new European regulatory regime.

ISDN became of interest to the Commission in the years 1982/83, when it initiated its first successful telecommunications initiatives. ISDN as a new network architecture was first discussed in international institutions, and from there it disseminated back to the national level. This was successful, because a small group of centrally-located actors was participating in CCITT, in the Confederation of European Posts and Telecommunications (CEPT), at the national level and later at the EC level, to promote ISDN. The justification for the CEC's new activities in telecommunications at the beginning remained within the realm of the traditional arguments legitimising national monopolies. The Commission stressed the important economic and social importance of a new telecommunications infrastructure, and the eminent importance of the telecommunications industry for industrial-policy purposes. Implicitly it was agreed that telecommunications was still a natural monopoly.

Based on cooperative talks between the national administrations, the Euro-ISDN project was born. This entailed the commitment to develop a pan-European network with a specified set of services to be offered Europe-wide and an interconnection of national networks. The old network operators, equipment manufacturers and the CEC were responsible for the details of this initiative. Parallel to these talks, coordination efforts at CEPT were continuing. The need to create a new European regulatory institution was discussed, but no firm action was considered to be necessary at this point, because of the obvious consensus. The common development of a telecommunications infrastructure was considered to be an innovative endeavour, but it was not considered to harbour any serious problems – in spite of the fact that the EC's ISDN programme went far beyond what the individual national administrations had planned with respect to ISDN. At this point, some network operators did not have any ISDN intentions at all. The EC's ISDN initiative was developed, however, by a group of dedicated ISDN experts, who perceived the rare opportunity to push through Euro-strategies with the help of an overall political climate that was favourable to such initiatives. Therefore, to the surprise of some of the actors the ISDN initiative was passed by the Council of Ministers in 1986.

In the same year the new Directorate General, Information Technology, Telecommunications and Innovation (DG XIII) was formed, which had a promising start with a variety of initiatives and programmes. An orientation towards industrial-policy aims still dominated these early years. Since the publication of the Green Paper, however, a new strategic orientation became prevalent: the perspective of the opening up of markets and deregulation. The Green Paper demanded a dismantling of the old telecommunications monopolies, especially a separation of regulatory and operational functions, as well as a separation of the network operators from other PTT enterprises. Finally, an opening up of procurement markets for equipment and service providers in all of Europe was foreseen. This meant that equipment and services should not only be offered Europe-wide, but also under the same conditions.

The network monopoly as such was not yet directly attacked – in spite of the fact that it was clear that the preservation of telecommunications monopolies was no longer part of the philosophy of the CEC. This can also be partially interpreted as a concession to the ISDN planners, because it was obvious that deregulated and possibly privatised network operators would hardly have the resources as well as the interest to provide the huge investments necessary for installing a universal pan-European network. If the CEC did not want to give up its ISDN plans, it had to guarantee the survival of the monopolies at least for a certain time. The original group of experts, who had advised the CEC on ISDN, continued to exist but it lost its strategic importance for the CEC.

The ISDN initiative was only one step in the continuing attempts by the CEC to acquire new competencies in telecommunications. Meanwhile, all member states have implemented the main aims of the Green Paper and have transformed them into national regulations. For ISDN this meant a slowing down of its introduction. The international standardisation in the old institutions of CEPT and CCITT nearly came to a halt. In some sensitive areas a policy of obstruction was conducted. The American Department of Commerce, for example, rejected the 'European' ISDN plans, because in their mind they signified a strengthening of the monopoly network operators of Europe. On the other hand, the territorially and nationally-based standardisation institutions, like CCITT and CEPT, were no longer representative of the economic and political environment.

At CCITT as well as in CEPT there were national delegations, which represented the interests of the network operators. With the separation of regulatory and operational functions in the network domain and the merging of computer and telecommunications networks, as well as the real existence of competing and private network operators, the old main actors lost their central position. In order to develop reliable standards and regulations, the group of people involved in negotiations had to be enlarged, new decision-making procedures had to be devised and new institutions created. The European Telecommunications Standards Institute (ETSI) seeks the widest possible market participation during the development of standards to ensure that the defined version meets real market demands and, therefore, has some chance of actually being implemented.

The old network operators began to rethink their network plans, because they were confronted with a changed situation in the market and new competitors. They had to cope with new and rival network concepts and a loss of dominance in the national market. The close cooperation with the national producers also came under pressure, because the latter had to orient themselves in the world market and smaller corporations were, to a large extent, unsuccessfully fighting for their survival. Long-term projects like ISDN were therefore handled with considerable caution. The slowing down of standardisation and the uncertain interest of the main actors reinforced themselves, and led to a dramatic deflation of expectations about ISDN. Industry developed equipment hesitantly and the network operators prolonged their plans.

Therefore, the EC's ISDN plans had to be adapted to the changing circumstances. With the Green Paper and the aggressively-pursued intention to push through its main aims, even against the resistance of the member states, the EC had established itself as a new decisive actor. This happened to the surprise of many of the national actors. They had allowed the CEC to become active in the field of telecommunications, because they did not perceive it as a threat to their status and, based on experiences from the past, were convinced that the Commission would fail. Towards the end of the eighties the situation had totally changed: where once the CEC was begging the national actors for cooperation, the CEC was now setting the agenda and putting the national actors under pressure. To give an example from the recent past, the Commission issued a review document on 21 October 1992, in which it listed the policy options for the future of telecommunications policy. At the same time as the Commission was asking for the opinions of the national actors, it also made clear what option it would pursue in the future anyway. The CEC is helped in pursuing its line of action by the European Court of Justice, which repeatedly has supported the Commission in its actions against the member states. The European Court of Justice's decision of late 1992 again confirmed the use by the CEC of Article 90 (3) procedure in the telecommunications-service case. The Court rejected arguments by Belgium, France, Italy and Spain that the Commission had overstepped its powers by issuing the Directive under Article 90 requiring member states to open their markets to certain telecommunications services.

On the national level therefore, a reregulation was going on, on the basis of EC recommendations, directives and recommendations. The changing role of the EC can also be seen when the role of the Commission in the ISDN process after 1986 is looked at. The activities of the Commission can be grouped in three areas: the reorganisation of standardisation; the strengthening of international regulation; and the extension of the actor network.

In the Green Paper, the CEC had already demanded that the process of standardisation should be thoroughly reorganised. The recommendations issued by CCITT would be too imprecise and the procedures too lengthy and untimely. Also, European institutions like CEPT could not fulfil the expectations of the Commission. A newly-created institution, the European Telecommunications Standards Institute (ETSI), was required to develop European standards, that not only took into consideration the needs of the old network operators, but also the aims of interconnected private and public networks, the opening of the closed national markets and the Europe-wide availability of equipment and services. This required cooperation between the old industrial standardisation institutions of the computer industry and telecommunications as well as the inclusion of new actors.

In the early eighties standardisation was considered to be a process that would not involve serious difficulties and, in 1982, industry and network operators agreed that international standards would be available in three to four years. However, the process of standardisation is still not finished and ETSI is looking for ever bigger meeting rooms to assemble all the people interested in the ISDN standardisation process. At this point it is very telling to mention that the ISDN standardisation work at ETSI is now chaired by a representative of a corporation that equips private networks.

In addition to the ISDN recommendation, a series of regulatory initiatives were started in order to speed up the success of ISDN. In this respect the open-network provision (ONP) has to be mentioned as well as the data-protection initiatives. Both ONP and the Europe-wide data-protection rules create new European law. ONP formulated a straightforward rejection of all proprietary networks and the standards governing them. ONP demands that networks have to be open to third parties and that the definition of technical interfaces, user conditions, tariff principles, etc, must be made public. Furthermore, ONP defines the boundaries between monopoly and competitive services.

Finally, the EU supports the creation of new European interest groups, which support the aims of the CEC. In the case of ISDN, the European ISDN User Forum (EIUF) has to be mentioned. The Forum organises user interests, but it is open to all parties interested in ISDN. The EIUF has privileged access to the CEC and its demands are routinely considered in the decision-making process.

Table 6. ETSI members

Membership Category	%
Manufacturers	61.5
Public network operators	14.0
Administrations	10.0
Users	7.8
Research bodies and others	6.5
Total number of full members	309

Source: ETSI April 1992.

The result of the CEC's involvement in network planning is obviously a mixed one. Just like the national planners, the CEC had to give up its initial bold aims. But unlike the national actors, the CEC established itself as a new and powerful actor in the domain of telecommunications policy.

The importance of the CEC can also be appreciated if one considers that a general ISDN availability before 1995 is only officially planned for EU countries plus the Nordic countries, Singapore, Taiwan and Switzerland [18]. Furthermore, the widespread interpretation that the CEC's ISDN initiative was propagated in order to get a minimum agreement between network operators after things had already gone wrong, is not adequate. The plans of the CEC and its experts were ready in 1982/83 and agreed upon by the experts. That the recommendation only went into effect in late 1986 is not the fault of the Commission. Clearly the 1986 recommendation – which is practically identical with the 1982/83 statements – would have looked different if it had been made on the basis of already vastly diverging ISDN schemes [19].

4. The future status of ISDN

The interesting question to ask now is, what will be the future of ISDN. As mentioned at the beginning, a dominant mood among telecommunications experts at the moment seems to be that the future of ISDN is already over. This is judged primarily on the basis of the original ISDN plans and the factors that have driven ISDN introduction so far. But as it has been pointed out, while the original ISDN plans have been given up, ISDN stays on. Furthermore, the complex causal configuration, which led to the introduction of ISDN, might say nothing about its eventual success. It is obvious, however, that ISDN as a strategic issue related to political strategies and interests is sunk. The Commission of the European Communities is still the main actor, and is trying to pursue a consistent policy strategy related to ISDN. This was supposed to reach a climax at the end of 1993, with a widely-publicised, multi-site presentation of ISDN (Eurie '93). But critical comments on the success of this event abound [20].

The Chairman of the Council of the International Telecommunications Users Group, William M. Coopman Jr, recently said: "If we are still waiting for SS7 and ISDN, should I be getting excited about Frame Relay, ATM, and some of the other megabit bandwidth services being touted? Why, after all these years, are the 'basic' services not available? Maybe before the carriers press too far into the future, they need to take stock of where they are in the present with their 'basic' offerings." [21]

The immediate policy problem is that ISDN was initially a strategic initiative of like-minded actors, that was supposed to suit the interests of those actors engaged in the process. The strategic element has got lost. ISDN still hovers around, but why and what for are the questions to be answered. The second problem is related but separate: ISDN has been discussed as an infrastructure initiative. In the back of people's minds was the thought that infrastructure, the speed and the volume of data to be transmitted over lines was a bottleneck problem. But nowadays hardly anyone thinks that the infrastructure as such is a problem. The problem will rather be, who will actually need and use all the bandwidths and all the possible networks on offer. The availability of infrastructure as such, therefore, is no longer a substantial issue, but rather what you want to do with the network, the services offered, etc. In this context, there is a new situation for the old network operators, who offer infrastructure for which there is no real need, and also for policy planners, who will have to redirect their attention to regulative issues dealing with the use of the network. In the western world the infrastructure argument, in spite of the fact that it is being reused in the broadband debate, has had no significant impact on economic activities in the ways perceived by the original advocates. Ten years ago, people claimed that without ISDN, regions, countries and the EC as a whole would lag behind in economic competition and welfare. Today the wide-ranging assessments about the importance of infrastructure have largely been abandoned. More specifically, the question to be asked is, for what purposes and at what prices will ISDN be a reasonable choice among the whole array of possibilities on offer at the moment.

Nevertheless, it is difficult to make a general statement about the future of ISDN, especially if one considers that real implementation in many countries has only just started. Based on the data, it is estimated that at the moment there are 44 countries that have substantial ISDN plans and the numbers are increasing. It will soon be a service offered all over Western Europe, hopefully with common inter-operability standards.

It will also be a service offered primarily by the old network operators. And it will remain an instrument to target the small and medium (business) users, who are satisfied with more standardised solutions. The old network operators hardly have the flexibility and the competence to cope with the very sophisticated and differentiated needs of large users. They will cater for some, but since all major network operators are after the same few big corporations, the gains to be made here will not be substantial. If more intelligently offered, ISDN may provide a means to cater for the needs of groups that are harder to penetrate by outside and new competitors. In this sense, ISDN might prove to be an important asset especially for network operators, who concentrate on their home markets and cannot be that aggressive in international markets. Old network operators are faced with a situation in which they have to endure a shrinking market share. They will have to reorient themselves. Not every operator can be a big player in the world market and some do not even intend to be one. Thus the network operators should go aggressively into the services and multimedia market. The question is: do they have the guts and the resources to do this. ISDN, which has so often been explained as a defensive move of the telecommunications industry against the onslaught of the computer industry, could now be interpreted as a strategic instrument of nationally-oriented network operators against the influence of international carriers.

Timms [22] has recently stated that ISDN will be the 'network of choice' for business communications during much of the 1990s. This seems to be supported by the difficulties of getting cheap and reliable broadband communications off the ground. In fact in some countries, ISDN has recently experienced an unpredicted boom. Even the optimistic estimates by Ovum [23] for the ISDN market in Germany, for example, have proved to be off the mark. The network operators themselves are surprised by the ISDN success after the doomsday prophecies of some years ago. Deutsche Telekom has reduced its investment in ISDN and customers now have to cope with long waiting periods for connection to the network.

But N-ISDN has also a lesson to offer for the B-ISDN process. With most of the network monopolies gone or to be discarded in the near future and the internationalisation of markets and networks, factors different from those in the N-ISDN case will be important for success. N-ISDN and B-ISDN share a problem in that economically it will take considerable time before it makes sense to implement these technologies on a larger scale. Therefore, new types of coalitions and policies will have to be developed to achieve early B-ISDN deployment. Both the Gore-Clinton initiative and the European RACE programme intend to cope with this situation. But obviously both approaches also have serious difficulties in getting people to agree on substantial issues like: who controls, who pays for what to whom, what is the exact role of governments in the process, who will be the eventual users, etc. ISDN was such an inspiring plan, because it had a strong and consistent support coalition: monopoly network operators, preferred equipment suppliers, national governments and the CEC. The process was supported by a tightly-knit, knowledgeable community of technical experts. All these things are lacking for B-ISDN, or more generally in the case of most broadband plans. Thus, ISDN might do better in the end than many expect.

ISDN started out in the early eighties to become something big and beautiful. It has ended up being something small and mysterious – but small might be beautiful as well.

Appendix

Table 7. ISDN cases

Country	A	B	C	Y
Germany	1	1	1	1
France	1	1	1	1
Belgium	1	1	0	1
USA	0	0	1	1
Japan	1	0	1	1
Canada	0	0	1	1
Switzerland	1	1	0	1
Australia	1	1	0	1
Singapore	1	1	0	1
Denmark	0	1	0	0
UK	0	0	0	0
Netherlands	1	1	0	0
Italy	1	0	0	0
Norway	0	1	0	0
Spain	1	0	0	0

Sources: DIW [13]; OECD [24] [25] [26]; Frensch [27]

Table 8. Top ten telecommunications equipment manufacturers 1980 and 1990

Manufacturer	Rank in 1980	Sales in billions US$	Rank in 1990	Sales in billions US$
Western Electric (USA)	1	12.032	–	–
ITT (USA)	2	6.041	–	–
Siemens (Germany)	3	5.054	3	9.864
Ericsson (Sweden)	4	2.878	6	7.121
GTE (USA)	5	2.199	–	–
Northern Telecom (Canada)	6	1.758	5	7.685
NEC (Japan)	7	1.448	4	7.732
Philips (Netherlands)	8	1.285	11	2.077
CGE (France)	9	1.919	–	–
Thomson Brandt (France)	10	1.630	–	–
AT&T (USA)	–	–	1	13.315
Alcatel (France)	–	–	2	11.581
Bosch (Germany)	–	–	7	4.455
Motorola (US)	–	–	8	3.629
Fujitsu (Japan)	–	–	9	2.723

Source: OECD [28] [29]

Notes and references

[1] Tony Newstead, ISDN: a solution in search of a problem? *Telecommunications Policy*, October 1986.

[2] Leonard R. Sussman, Power, the Press and the Technology of Freedom: the coming age of ISDN. Freedom House, New York, 1989.

[3] Most calculations on user-needs for broadband solutions do not consider that there might be more than one network provider. (See for example, Hagenhaus and von der Straten.) [30]

[4] John M. Griffiths, ISDN Explained: worldwide network and applications technology. John Wiley and Sons, Chichester, 1990.

[5] CCITT: Comité Consultatif International de Télégraphique et de Téléphonique.

[6] International Telecommunications Union, CCITT, Data Communication Networks Services and Facilities, Recommendations X.1–X.15. CCITT Red Book, Vol. VIII, Facsimile VIII.2, 1985, pp 45, 50, 52.

[7] While some see ISDN especially as an European project with a strong role for the CEC (Darmaros) [31], others stress the national side (Kahl) [32] and deny that the CEC had an important impact on the planning process. See also Fuchs [33] [34].

[8] This statement is not very illuminating, because most developments in the hardcore sector of telecommunications are technology driven. Good and more recent examples are the broadband and asynchronous transfer mode plans. In these cases – just as with ISDN – there is only an assumed but not yet real demand. The important players, however, feel obliged to react and participate, for example, in the race towards fibre in the loop, to cope with a demand that may arise – which, of course, could also be awakened by the provision of the specific infrastructure. Besides, on the surface ISDN was at least as much the result of economic thinking, that was not oriented towards short-term benefit calculations, but was in the tradition of the national network planning oriented towards a supposed 'optimum'. In 1986, Noam [35] stated that he could not find any reliable data concerning a demand for ISDN. Three years later, in 1989, there were many studies on the demand development available which, however, contradicted themselves in a way that made them nearly useless.

[9] The social science literature is looked at primarily, not only because the author is a social scientist by profession. Economy and technology-oriented literature emphasises both the importance of technological developments and (economic) conditions favouring or preventing the success of ISDN. They under-rate the importance of the fact that ISDN clearly has been (especially in the early years) a strategic political issue for which only technological and economic arguments were used.

[10] A more in-depth critical evaluation of these three interpretations is provided by Fuchs [36].

[11] Charles C. Ragin, The Comparative Method. Moving Beyond Qualitative and Quantitative Strategies. University of California Press, Berkeley, 1987.

[12] This implies that pilot installations and technical trials are not included in the argument. However, offers that are only similar to the standardised ISDN version, such as British Telecom's precursor solutions, are also not considered.

[13] DIW (Deutsches Institut für Wirtschaftsforschung), The Spectrum of Alternative Market Configurations in European Telecommunications. Part I: European Telecommunication Organisations. Selbstverlag, Berlin, 1988.

[14] Data not available in the sources listed have been added, based on the author's research.

[15] Note that in Table 2 only commercially-available ISDN accesses are counted, while Table 1 considers in a more general manner, all systems in operation including trial installations and pilot projects.

[16] Upper case letters signify existence and lower case letters non-existence.

[17] Primary-rate ISDN provides thirty 64 kbit/s channels – the bearer or 'B'-channels. Hence it is often referred to as 30 B+D. Basic-rate access offers two B-channels which may be used for voice, data, text or image, plus a single 16 kbit/s D-channel. The service, 2B + D, is well established with a CCITT-ratified recommendation I.420. Primary-rate ISDN is not that fully standardised yet.

[18] Slowenia has also stated its intention to offer universal accessibility to ISDN. But the first experimental switches will only start operating some time in 1994. The year stated for Japan is 1996.

[19] Recently Warwick [37] among many others, has strongly advocated the 'failure' view.

[20] See *Communications Week International*, 11 October 1993, pp 1 and 4.

[21] Coopman's editorial comment in *International Telecommunications Users Group News*, October 1993, p 2.

[22] Stephen Timms, The world scene and will ISDN be overtaken? Institute of Electrical Engineers Colloquium on ISDN, London, 11 November 1991.

[23] Ovum, Cost-based tariffs for data services: position and trends in Europe. London, 1989.

[24] OECD, Major R&D Programmes for Information Technology. OECD, Paris, 1989.

[25] OECD, Performance Indicators for Public Telecommunications Operators. OECD, Paris, 1990.

[26] OECD, Communications Outlook 1990. OECD, Paris, 1990.

[27] Karl J. Frensch, ISDN -Einführung- und Marktperspektiven im internationalen Vergleich. Working paper, Munich, August 1990.

[28] OECD, Telecommunications. Pressures and Policies for Change. OECD, Paris, 1983.

[29] OECD, Information Technology Outlook 1993. OECD, Paris, 1993.

[30] Lutz Hagenhaus and Gernot von der Straten, Broadband Networks: will there be users? Determining how many users will subscribe to broadband services in industrialized countries, *Telephone Engineer and Management*, Vol. 97, April 15 1993.

[31] T. Darmaros, Beyond the Sales Pitch: realising ISDN in the US, Japan and Europe. Centre for Information and Communication Technologies, Working Paper Series No. 5, April 1989.

[32] Peter Kahl (Ed.), ISDN. Das neue Fernmeldenetz der Deutschen Bundespost TELEKOM. R. v. Decker's Verlag, Heidelberg, 1990.

[33] Gerhard Fuchs, The Politics of Pan-European Telecommunications Network Development, *Journal of European Integration*, Vol. 16, 1992.

[34] Gerhard Fuchs, Integrated Services Digital Network: 'The Telecommunications Highway for Europe After 1992' or 'Paving a Dead End Street?' The Politics of Pan-European Telecommunications Network Development. Discussion Paper 6/93. Max-Planck Institute for Social Research, Cologne, 1993.

[35] Eli M. Noam, The Political Economy of ISDN: European network integration vs. American system fragmentation. Paper presented at the XIV Annual Telecommunications Policy Research Conference. Airlie, Virginia, April 1986.

[36] Gerhard Fuchs, ISDN - Einige Anmerkungen zur (wissenschaftlichen) Diskussion über die Entwicklung von Telekommunikationsnetzen. Working paper, Cologne, 1991.

[37] Martyn Warwick, ISDN: a technology still in the wilderness? *Communications International*, April 1993.

Telecommunication: Exploring Competition
H. Williams and M. Bonnan (Eds.)
IOS Press, 1994

Chapter 5

Abolishing VANS? valued-added network services vs public switched services in electronic data interchange communications

Anders Henten
Institute of Social Sciences, Technical University of Denmark
and
Steen Ottosen
Telicon, Denmark

Abstract: This paper discusses the future of value-added network services in the light of political/ideological, economic and technological developments, and examines the case of electronic data interchange communications running on value-added network services and public switched integrated services digital networks respectively. Our conclusion is that electronic data interchange communication does not necessarily need any value-added network services intermediary between two communicating parties and that integrated services digital networks are a far cheaper solution than value-added network services.

1. Introduction

1.1. Defining VANS

The emergence of value-added network services (VANS) is closely related to the liberalisation of telecommunications during the last 10–20 years. VANS are, furthermore, connected to the digitalisation of telecommunications and the growing number of new services building upon the convergence of data processing and telecommunications. In some accounts the term VANS, on the basis of these facts, seems to acquire an almost mythical character as the ultimate manifestation of liberalisation, and new telecommunications services are being portrayed as the prerogative of VANS.

However, new services are not necessarily the birthright of VANS, and further liberalisation may even turn out to become an increasing problem for VANS as a specific category of network and service providers. Advanced services may just as well run on public networks, and the present type of dichotomy between Public Telecommunications Operator (PTO) networks and private VANS networks will disappear – or at least be blurred – when not only services but also networks become a liberalised field [1].

Defining VANS is rather difficult both because the term VANS has somewhat diverse meanings in different countries and because the whole field is extremely volatile. At least three definitional aspects must be dealt with: there is a legal and regulatory dimension, a technological dimension and a third dimension concerning actual VANS companies.

The first dimension deals with the attempts – first in the USA and later on in the UK and elsewhere – to determine a regulatory dividing line between operators acting in the basic and the enhanced area respectively (using the terminology from the 2nd Computer Enquiry in the USA). In the USA, the aim of establishing such a dividing line has been to prevent AT&T (American Telephone and Telegraph) and the regional Bell operating companies from entering the enhanced field. In Europe, the aim has been the opposite, namely to protect the basic field from new service operators.

The second dimension regards the technological side of the same problem, viz. to define the different layers in telecommunications (bearer services and service applications). The development of new service applications has emphasised the importance of establishing technologically independent layers in the communications system as in the open systems interconnection (OSI) model.

The third dimension deals with the actual companies operating in the VANS area. Their primary fields of operation may change (for example, from protocol conversion to specific service applications). They, however, continue to exist as telecommunications operators.

The regulatory dimension changes over time, influenced by political, ideological, economic and technological developments. The technological dimension has a more stable character – but only to a certain extent. Technological developments continuously break down the barriers between basic and enhanced services. The company dimension is, in a sense, the expression of the legal and technological opportunities for delivering value-added network services.

Clearly, the political regulatory dimension is the driving force in the present tendency to break down the barriers between VANS and common carriers. This does not mean, however, that existing VANS firms will cease to operate, but they will not constitute a separate category of network and service providers. Therefore, new lines of demarcation between different types of telecommunications operators must be developed.

1.2. Aim of the paper

Electronic data interchange (EDI) is one of the new data-communications services emerging on the basis of the integration of data processing and telecommunications. In some business sectors, EDI has already reached a take-off stage and will, in the near future, become a widespread service substituting for paper-based procedures. For those VANS companies that have EDI as a major service area, future prospects should on this background seem fair. But EDI services do not necessarily have to pass a VANS operator. EDI services could just as well be delivered on public networks which would pull the rug from under the feet of a number of VANS companies.

In this paper, we set out a presentation of the inter-relationship between EDI and VANS. The primary purpose of this exercise is to analyse whether there are other means of transmission that are just as efficient – or even more efficient – and at the same time more cost-effective than the 'VANS solution'. The secondary and broader purpose is to consider this question in the light of the continuous cycle of technological fluctuations between innovation and standardisation.

These two issues are joined together in the question of public networks vs private networks. Are networks conclusively tipping away from being public (as suggested, for example, by Eli Noam [2]) or are they shifting back again after a period with private networking experiments (as assumed, for example, by the CEC PACE '92 study [3])? Or is there a third possibility where the strict dichotomy between public and private networks disappears and more hybrid forms develop? [4]

1.3. Public/private

In order to be able to answer this question, it is necessary to define the concepts 'public' and 'private' in relation to telecommunications networks, and it must be acknowledged that any such definition has to be understood in a specific historical context – although not too specific for generalisations to be elaborated.

We agree with the proposition made by Áine NíShúilleabháin [4] that access (from general to dedicated) is the fundamental definitional criterion when discussing public vs private networks. However, we do not agree that it is a sufficient criterion. The access criterion may be sufficient in a situation with a clear-cut difference between public networks with general access (for example, public switched telephone networks (PSTN)) and private networks with dedicated access (for example, leased lines). The question is where to place VANS? Are they private networks because they operate on leased lines? Or are they public networks because there is general access?

To answer that question, at least one additional dimension should be added to the concept of public service, namely, the concessionary status of operators and the licensing conditions, including the responsibilities of operators towards publicly-decided goals and means. Such responsibilities could include provisions for universality and pricing, with the aim of ensuring the delivery of services even when they are not profitable. The decisive point is the purpose behind the delivery of services, ie, whether there is a broader societal purpose transcending the aim of sheer profitability.

Today, many PTOs are standing in the middle of the road trying to fulfil both requirements: profitability and socio-political goals. They often experience such socio-political goals as a burden that they would wish to shed and, presumably, a number of these operators will eventually be permitted to 'go free'. They will then be crossing the line between public and private operators – even if publicly owned. Ownership is not a decisive point – although ownership and purpose normally will be inter-related to some extent.

Following this extended definition of public services, VANS are clearly not public networks. Integrated services digital network (ISDN), on the other hand, is a clear-cut example of a public network. It may be true that technically speaking (digitisation from end-to-end, 2 x 64 kbit/s + 16 kbit/s [5], service integration), ISDN is not necessarily a public network service. However, taking into consideration the way this network has been conceived and developed, it is a very good example of an advanced public network.

In this paper, we will focus primarily on ISDN and VANS as different modes of transmission for EDI communications. We are well aware of the fact that VANS operators may have ISDN in their service portfolio and that public ISDN lines may be used to communicate with a VANS computer, but from a customer's point of view, VANS, ISDN and other network services are alternative modes of transmission where quality, service features and prices are the basis of choice between different solutions.

2. Electronic data interchange

2.1. Standardisation

Electronic data interchange (EDI) is defined as electronic transfer of structured data by means of standardised communications applications from one computer to another computer. For our purpose, the interesting part of the definition deals with the standardisation of communications applications. It is, of course, also vital that data from one computer is structured in such a way that it can immediately be processed in a receiving computer without any human validation, which is the field of EDIFACT (electronic data interchange for administration, commerce and transport) message development. Message development is not, however, the object of our interest here. That area only becomes interesting because EDIFACT is extremely important for EDI communications to gain widespread dissemination.

From a technical point of view, there are a number of barriers to EDI communications taking place, of which communications protocols and structuring of data are some of the most important. With EDIFACT becoming more common, the basis for a broader implementation of EDI will be improved, which will then call for generalised solutions to the transmission problem. Technically speaking, there is no necessary connection between the standardisation of transmission protocols and the harmonisation of EDI messages, although they are both necessary preconditions for EDI to become a widely-used service.

2.2. *Market potentials*

The potential number of documents that can be transferred by EDI must be based on the number of paper documents used in today's business. Although the introduction of EDI may in itself change the number of documents when business procedures are changed, we expect, in the following, that these changes are neutral.

Previous research in Denmark has estimated a total average of approximately 300 trade-related documents per one million Dkr turnover in the trade and manufacturing sectors. There are no similar figures for other important sectors like finance, insurance and the whole public administration.

The total turnover in the trade and manufacturing sectors of Denmark equals 1,000 billion Dkr. Based on these figures, some 300 million trade-related documents circulate per year in these two sectors in Denmark. If we include other important sectors, there must be at least one billion trade documents circulating per year in Denmark. Extrapolated to the whole EU area, this will make some 50 billion trade documents each year.

2.3. *Paper vs electronic communication*

If electronic communication is to become a real alternative to paper-based transmission, it is imperative to find the cheapest, most simple and reliable way of transferring electronic documents. At present, transmission of trade documents such as invoices, payment orders, orders, etc, is predominantly in the form of paper, envelopes and stamps, despite the vast amount of work that has been done in order to switch to purely electronic transmission of trade documents.

The trend towards standardisation of telecommunications services establishes an opening for new and better ways of communicating directly between any two parties. Electronic documents have a lot of built-in advantages: they are fast to transmit; there is no re-entering of data (no errors); and documents/data can be integrated with the most commonly-used administrative programmes in companies. So why is most transmission of trade documents still in paper form?

Part of the answer is that it is too complicated and expensive to send the information by electronic data transmission. Therefore, a concept of 'direct EDI' communications must be implemented, where all electronic communications and security problems are handled as simply and cheaply as present technology allows.

This may be a controversial idea in some circles as there are many advanced commercial VANS solutions available on the market. It seems as if it is necessary for any electronic information to pass at least one extra computer system and one extra communications network to make the EDI transmission 'right'. If this continues, EDI will never be a real alternative to paper communications.

3. Different means of communication

EDI is not a new phenomenon. Long before the name was even 'invented', bilateral EDI communications were performed on leased lines. Later on, VANS companies came in as intermediaries between the companies needing EDI exchanges. The reason for this development being that bilateral EDI is a very inflexible solution. Companies have to lease a line for each business partner to take care of electronic communications. VANS companies enter the scene here, delivering a multilateral service by selling (value-added) capacity on their leased lines. But VANS companies are not necessary intermediates. EDI services can just as well run on public networks.

In the following, we will briefly look at six different means of communication:

* PSTN	* Leased lines	* VANS
* X.21		
* X.25		
* ISDN		

We have subdivided the different means of communication into three categories: public switched services, leased lines and VANS. These categories are, of course, different in nature: in the case of public switched services, we are dealing with four bearer services; and in the cases of leased lines and VANS, we are dealing with different ways of utilising such bearer services. However, from an end-user's point of view, all six solutions are alternatives to be compared with one another.

The *PSTN* solution involves data transmission on the traditional telephone network by means of a modem and dialling. This is both a flexible and cheap option when the amount of data is relatively small. With larger amounts of data, *leasing of lines* becomes relevant. But this is, as mentioned, a very inflexible solution as it only allows communication between two pre-defined subscribers. *X.21* is the name of the CCITT (Comité Consultatif International de Télégraphique et de Téléphonique) recommendation for circuit-switched data networks [6]. X.21 is normally used for transactions of very small data files because the time taken to establish a connection is very short and a new connection, therefore, can be established for each transaction keeping connection time down. *X.25* is a packet-switched network service which is being used because the efficient use of a circuit-switched connection often is very low. As with X.21, the purpose is, therefore, cost-effectiveness but the problem is that both these networks are too specialised for the larger number of potential subscribers to become connected to the service. *ISDN* is, in a way, an extended PSTN solution in terms of potential flexibility and generalisation. Furthermore, ISDN adds digitisation from end-to-end, higher capacity and service integration. The present problem with ISDN is its very low level of dissemination. The *VANS* solution is also very flexible, but at the same time, it is a very expensive solution.

In Table 1, we summarise some of the characteristics of the different means of communication mentioned. We are aware of the fact that such characterisations are problematic because they leave out important features. The table presents an overview of characteristics that are important when choosing a communications solution that matches the needs of an EDI user.

In Denmark, the most commonly-used communications solutions are PSTN, leased lines and VANS, but ISDN should become a realistic alternative in the future. In all developed countries, a lot of money and effort have been used in changing the telephone network from analog to digital. In fact, nearly all internal communications in the telephone systems are now purely digital, and ISDN extends this digitalisation of the telephone system the last stretch to the end-users via the normal installed copper wires.

Table 1. Comparison of six different means of communication

	PSTN	Leased lines	X.21	X.25	ISDN	VANS
Communi-cation	Multilateral	Bilateral	Multilateral	Multilateral	Multilateral	Multilateral
Service integration	Low	Low	Low	Low	High	High
Application protocol	No conversion	No conversion	No conversion	No conversion	No conversion	Protocol conversion
Speed	≤9.6 kbit/s (a)	Variable	≤9.6 kbit/s	≤64 kbit/s	≤2 x 64 kbit/s + 16 kbit/s (b)	Variable
Cost – fixed – variable	 Low Low	 High negligible	 High Low	 High Low	 High (c) Low	 High High

a: Modems with higher transmission capacity can be used on PSTN.
b: This is the European ISDN standard.
c: The reason for characterising ISDN fixed costs as high is that ISDN adaption equipment is still expensive.

For ISDN, there are comprehensive OSI standards that define in detail how computers are connected. This dramatic change in the telephone systems has produced some practical problems but it is absolutely the best network for directly communicating EDI. Furthermore, the ISDN network is the cheapest because the telephone companies *can* sell 2 x 64 kbit/s plus 1 x 16 kbit/s for exactly the same price as a normal telephone line – there are no extra costs for the telephone company. In fact, this is the pricing system for ISDN in Denmark.

One problem that ISDN cannot solve is protocol conversion. Conversion has been one of the strong points of VANS. The VANS computers have converted communications protocols and have in this way solved many communications problems between different EDI communicating parties. Protocol conversion has, as a consequence, been one of the more lucrative business area for VANS operators. The question is whether protocol conversion remains to be an actual problem.

This is a crucial point if ISDN is going to become a major means of transmission for EDI communications. ISDN can transmit different types of application services and it can also be the conveyor for other network services such as X.21. It cannot, however, convert from one communications standard to another. The communications standard has to be the same at both ends of the communication.

As standardisation goes along there is, on the one hand, a vast amount of more or less obsolete standards rarely being used and which will be extremely costly to convert. On the other hand, there is a small number of standards being used more and more widely (X.400, FTAM and TCP/IP [7]) and conversion, therefore, tends to lose its importance.

4. Costs of EDI communications

For the users, the price of the service is one of the most important parameters in the choice between competing systems. This, of course, presupposes that services are comparable in quality. With the decreasing importance of protocol conversion, ISDN and VANS solutions, in our opinion, become comparable.

In this section, we compare prices of transferring EDI documents on VANS and ISDN respectively. It should be noted that the VANS solution includes transmission on ISDN lines and that the difference, therefore, has nothing to do with the choice of bearer service but with the question of using an intermediary between any two communicating parties. If other bearer services are used, the difference in costs between a VANS solution and a direct EDI solution will be larger because VANS typically require twice as much communication time on the network: both from sender to VANS and from VANS to receiver. ISDN is the cheapest bearer service for EDI communications.

Electronic transfer of documents can be done through many types of physical and logical connections. In this comparison, only one set of presumptions is used, as the trend is the same no matter the size and type of the computer system. The presumptions made here are:

- ISDN network, basic rate (2 x 64 kbit/s + 1 x 16 kbit/s)
- minicomputer connection, Unix based
- depreciation of investments over two years
- average document size 3 kbit
- all documents are transmitted batch, not real-time

If we make the above presumptions, the cost-relationship will be approximately 1:2 in favour of direct EDI when transferring 1,000 documents and 1:5 when transferring 10,000 documents (see Figure 1). If documents are transmitted on-line/real-time, 4 Dkr per document – which is the extra cost of real-time/secured-transmission on VANS – would have to be added to the costs of a VANS transmission.

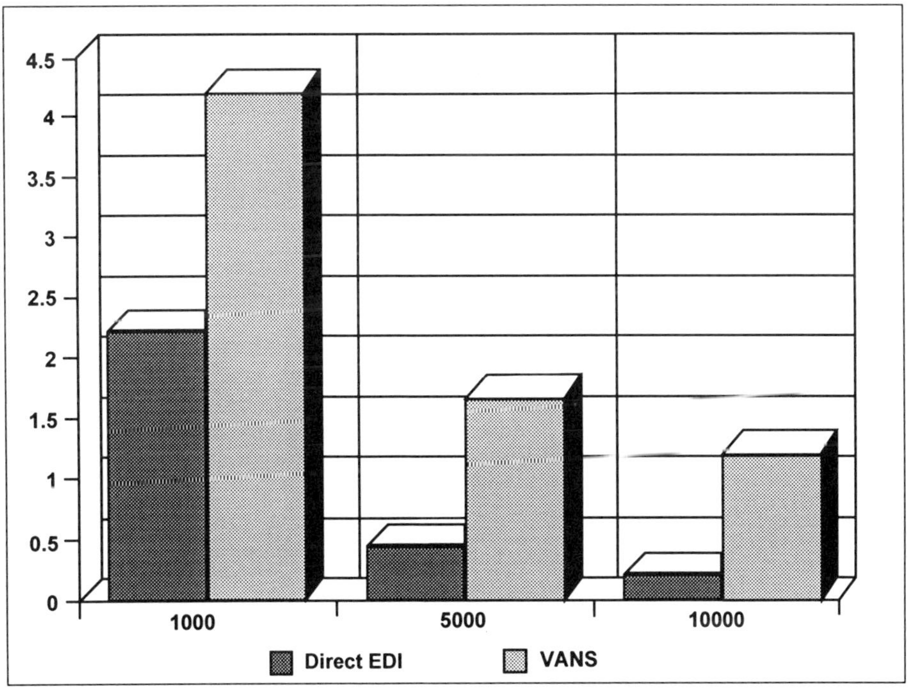

Figure 1: The costs of EDI transmission

Note: The X-axis represents the number of documents, while the Y-axis represents the costs per document in Dkr.

If we use these costs for EDI transmission via VANS or directly via ISDN, it is possible to calculate the potential impact on the national economy. The total number of documents per year has been calculated to be around one billion in Denmark. In a Danish context with many small and medium-sized enterprises, it is realistic to utilise the figure of 1,000 documents per firm per month as an approximation. If we assume that half of all documents are transmitted on-line, the total yearly national costs of a VANS solution would be 6–6.5 billion Dkr compared with 2–2.5 billion Dkr in the case of an ISDN solution.

The principal reason for this difference is that VANS are an expensive additional link in the communication between two parties. This primarily concerns the costs of computer time on the VANS computer and the necessary personnel to run a VANS firm. The costs of 'double transmission' (from sender to VANS and from VANS to sender) is only a minor additional cost.

5. Direct EDI

In this section, we give a more detailed presentation of the 'direct EDI' solution. The term direct EDI is being used to describe an EDI communications system running solely on public networks without any detours around VANS operators.

The most relevant transmission system to compare it with, if discussing a generalised system for interchanging documents, is in reality, the paper-based mail system. That is the system to substitute, and it is in relation to that system that different telecommunications solutions must measure their potentials and advantages.

Transmission of paper-based trade documents is almost exclusively the way of doing business today. Computer systems, programmes and internal routines are designed to support paper documents. This is no coincidence, because paper document-based systems have many well-known advantages. Therefore, it is important to carefully match (or improve) these advantages when electronic-transfer systems are designed.

These advantages can be summarised as follows:

1. Flexible and simple
2. Homogeneous communications
3. Standard format
4. Safe
5. Authorisation friendly
6. Original document for auditing
7. Legal evidence

1. *Flexible and simple*: as in the paper-based postal system, there is a need for simple addressing procedures in the case of electronic communications, and one can hardly think of a more simple and worldwide addressing system than the public switched telephone system.

2. *Homogenous communications*: as for the transfer of paper mail, a number of electronic networks are internationally connected. The most common of these networks is the telephone network available in all countries around the world. The telephone network could be used for direct EDI communications but, unfortunately, there is no common standard for the connection of digital computer equipment to this analog network. Digitisation from end-to-end will help solve this problem.

3. *Standard format*: a standard for the design of electronic trade documents is already being prepared by working groups under the aegis of United Nations Economic Commission for Europe. The EDIFACT standard defines the format and the syntax of any information type needed in a trade document. EDIFACT is developed and accepted worldwide and must be used as the EDI standard for directly-communicating EDI systems.

4. *Safe*: it is very difficult to replace the security routines used in today's paper-based documents. But encryption systems exist that ensure both digital signatures and encrypted-data transmission. From an objective point of view, these systems are far more secure than paper documents, but there is a lot of subjective fear because the technology is new.

5. *Authorisation friendly*: a document that only exists in electronic form must be sent for electronic authorisation in the organisation. This is, probably, the most expensive and difficult part of implementing EDI and calls for flexible communications systems.

6. *Original document for auditing*: electronic documents must be stored as genuine and original documents in only one version both by the sender and the receiver. This makes it possible to use the electronic documents for auditing. Here also, encryption and electronic signatures must be used.

7. *Legal evidence*: disputes concerning the transmission and the contents of electronic documents will arise when more and more documents are transmitted electronically. The systems used for encryption and digital signatures must have a quality that provides a very high degree of probability of evidence. Such systems already exist and are in use today at relatively low prices.

In our opinion, a 'direct EDI' solution can match all these essential requirements. With direct EDI, electronic documents are transmitted directly from sender to receiver through a public switched network using a normal telephone number for addressing. Potentially, everybody can contact everyone else without compromising security measures.

ISDN seems to be the appropriate conveyor solution for 'direct EDI'. The present problem is that in most countries there are very few ISDN customers. This problem is, however, not of any greater size than it is with the VANS solution. In both cases, EDI users will have to become subscribers to the respective systems.

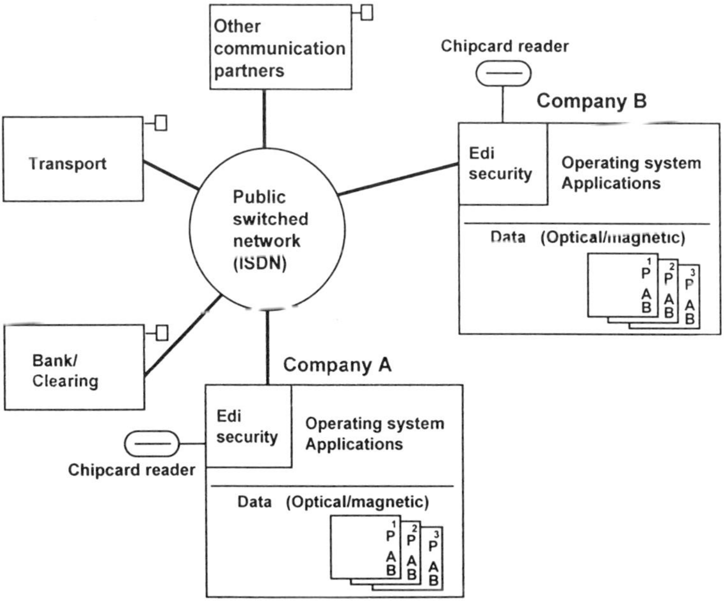

Figure 2: EDI on public switched ISDN

An ISDN-based system is presented in Figure 2. The figure shows two companies A and B connected to each other and to three additional business partners through a public switched ISDN. The EDI computers are standard computers equipped with ISDN communication cards, chipcard readers and optical/magnetic units. The security system is based on the use of chipcards. These cards contain small monochip computers that prevent the reading or changing of data. After transmission, the documents may be saved on, for example, an optical WORM (write once read many times) disc.

6. Conclusion

In this paper, we believe we have shown that EDI communications do not necessarily have to pass through a VANS operator. It is technically feasible to develop a flexible and generalised EDI system running on ISDN and it is much cheaper for the customers.

A crucial element in this line of argument is that a small number of communications protocols have emerged as the dominating ones (for example, X.400, FTAM). Although a vast number of different communications protocols continue to exist, a selection procedure has taken place both in official international organisations such as CCITT and the International Standards Organisation and in the 'natural selection process' outside the official standardisation bodies.

Liberalisation has given way to societal experiments with a multitude of different communications standards and solutions. This has been a necessary development – and still is. EDI is a very good example: it is not likely that the incumbent telecommunications operators would have contributed as positively to the development of EDI and communications standards as VANS operators have done, but it seems as if a stage has been reached where more global solutions should be developed.

In Denmark, EDI standardisation and adaption to international standards have taken place in sector-specific cooperative organisations in which not only users but also VANS operators have participated. This has, on the one hand, been a very slow and sometimes cumbersome way of doing things. On the other hand, it has secured a high degree of relevance of the standardisation work.

A higher level of coordination is now on the agenda. In Denmark, an EDI Council has been established with the purpose of coordinating EDI activities and in this way strengthening the implementation of EDI. In an inaugural statement, the EDI Council writes:

Danish business enterprises are, at present, in such a difficult situation that many firms do not have the strength to establish EDI solutions. The sector-specific initiatives that are established or are under establishment generally suffer from economic difficulties with the result that solutions come too slow or do not come at all. With the purpose of furthering the EDI development in Denmark and of strengthening the competitive edge of the individual firms, economic means should for a period of time be allocated for the improvement of the EDI work in Denmark. [8]

The primary focus of the EDI Council is on standardisation of EDI messages. We would like to add standardisation of the means of communication. If EDI is going to develop into a widely-used infrastructure for document transmission, EDI solutions must be both easy to implement and cheap. If not, smaller and medium-sized enterprises will stay out and as a consequence lose some of their competitiveness. The hub and spoke configuration or the VANS solution may be suitable for a period of time – although these systems often function on the premises of the larger corporations. There is no reason why more generalised network solutions should not be implemented when possible. This is the background for our proposition to advance the concept of 'direct EDI'.

VANS operators have been very active in the EDIFACT message-development work.

In combination with the fact that EDI is a value-added service, this has led to the erroneous conclusion that EDI is one of the prerogatives of VANS – an idea that we have rejected in this paper.

It is quite certain that some VANS operators will face economic difficulties if a 'direct EDI' solution gains ground. However, by no means all VANS operators have EDI as a major business area. In Denmark, the VANS operator danNet (a subsidiary of KTAS, Kjøbenhavns Telefon Aktie Selskab and IBM Denmark) has given a lot of attention to the EDI field. Internationally, GEISCO (General Electric Information Services) is one of the major EDI transmission networks.

Other network operators mainly concentrate on network management. TRANSPAC (the French packet-switching network) – often considered the biggest European VANS – is an outstanding example of such an operator. Also in the case of network management, the establishment of public switched high-capacity networks constitutes a problem. The 'VANS area' is, in other words, under attack from the very same technological development that was part of the foundation for the emergence of value-added network service operators.

This does not at all mean that the actual VANS firms will cease to exist. With the ongoing liberalisation of telecommunications, including the physical networks, there will be new business opportunities for such firms, although the reason for preserving a special category of network operators named VANS seems to vanish. The question mark in the title of the paper, 'Abolishing VANS?', may, therefore, be substituted with an affirmative mark, but only in the above sense.

The term VANS operator will lose meaning when VANS operators are permitted to offer pure network services (in Europe) and common carriers are permitted to offer value-added services (in the USA). The term VAS, meaning value-added service operators without their own network, will perhaps continue to have a meaningful content – no matter whether VAS firms are subsidiaries of network operators or independent firms. Value-added services, such as information services and electronic funds transfer systems, will exist in the future, offering their services on different networks.

In contradiction to the demarcation line between value-added networks and pure transmission networks, the line between public and private networks will continue to have an importance. In the old sense of the words 'public' and 'private', they tend to lose their meaning because networks that formerly had a more private character, in the present situation may be characterised as public on grounds of accessibility. In that sense the demarcation line between public and private is being blurred. But if an imposed commitment to politically-decided goals and means is added to the concept of public, then the terms have an operational content.

The traditional bargain between a telecommunications operator and the public authorities, trading monopoly rights for a commitment to universal service, is no longer possible or desirable. Monopoly rights will only be granted in special cases, but public authorities still have the possibility of exchanging certain advantages for a commitment to politically-defined goals.

This is not primarily a 'normative affair', ie, a question of whether public networks per se are 'better' than private networks. The question is whether is should be possible to set political goals in the field of telecommunications. We believe that it should be, and that one of the ways of doing this is to give certain advantages to telecommunications firms that are prepared to meet specified requirements.

One of the reasons for conducting such a policy is that not all aims of society will necessarily be met by way of private interests alone. Another reason – which we have documented in this paper – is, that what is profitable for a category of private firms is not always profitable from a societal point of view. In the case of EDI, our calculations show that an all-electronic EDI system conducted by VANS operators would cost 6–6.5 billion Dkr per year. A similar all-electronic system, running on ISDN, would cost 2–2.5 billion Dkr per year.

Notes and references

[1] In the USA, network provision has been an open field for years. It is, therefore, characteristic that the term VANS, in the USA, covers not only value-added services but also simple data-transmission services, which is not the case in Europe.

[2] Eli Noam, Telecommunications in Europe. Oxford University Press, Oxford, 1992, pp 26–42.

[3] PACE '92, Perspectives for Advanced Communications for Europe: executive summary. Commission of the European Communities, Brussels, 1992.

[4] This is, for example, the position of Áine NíShúilleabháin, Strategic Responses by Public Network Operators to Private Networking in Europe. Columbia Institute for Tele-Information, New York, Working Paper Series, 1992.

[5] The European ISDN standard.

[6] In some countries, X.21 is not offered as a separate physical network but as a service on ISDN.

[7] FTAM = file transfer, access and management
TCP/IP = transmission-control protocol/Internet protocol

[8] Danish EDI Council, EDI – Electronic Data Interchange, 1993.

Telecommunication: Exploring Competition
H. Williams and M. Borman (Eds.)
IOS Press, 1994

Chapter 6

Universal Service: key issues for the next decade

Jacques Arlandis
IDATE, Montpellier, France

Abstract: This paper examines the problems which surround the definition of universal service in a newly competitive and regulated telecommunications sector. The traditional definition, which links the concept to monopoly provision, is shown to be relatively simplistic in the context of the current telecommunications sector. This new context justifies a review of the concept of universal service and, as shown in this paper, universal service and competition are intimately linked. Redefining universal service can only occur alongside a critical evaluation of both competition and industrial dynamics.

1. Introduction

It was with a certain short-sightedness that the debate on the review of European telecommunications policies was embarked upon in autumn 1992. But under the pressure of national parties, it was broadened from its initial focus on market liberalisation, so as to explicitly take into account the universal-service issue. This led the Commission of the European Communities to reformulate the stakes in a slightly different way from what it had initially specified in its draft document of October 1992, in which it set out scenarios delineating four possible European market structures.

The debate was shelved by the adoption of the resolution of the Council of Ministers of 22 July 1993. One statement sums up the direction this policy will henceforth take: 'one of the principal objectives of Community Telecommunications policy, is the liberalisation of all voice telephony services whilst maintaining universal services' (in Council Resolution 93/C213/01).

The Commission's draft document was effectively singularly taciturn on the problem of maintaining or developing universal service, or both, as an element of reference in public policies in Europe. National policy makers, supported by public operators, argued against this weakness. To calm things down and bring some order (and consensus) into the debate, the Commission then produced a specific communication [1] which is currently being discussed: 'Developing universal service for telecommunications in a competitive environment'.

The wording is stimulating. It could provide grounds for a more problematic reformulation: *in what form will universal service* (the foundation of fifty years of telecommunications growth in Europe and elsewhere), *survive the introduction of competition in the telecommunications world?*

We shall try to contribute to this debate, more particularly on the rationale which underlies the evolution of the very concept of universal service.

The analysis has thus been organised in three stages:

Section 2. a discussion of the coverage of the universal-service, public-service and common-carrier concepts.

Section 3. a deliberation based on four cases, relevant to our key question: the integrated services digital network (ISDN) case, the Internet case, the Iridium case and the personal-

communications network/personal-communications service (PCN/PCS) case,
Section 4. an identification of the issues that should be on the next agenda dealing with universal-service policies.

This paper argues that universal service is a key techno-economic construct, embedded in the monopoly/competition bargaining process. As the new industrial dynamic requires a change in market structures, the concept of universal service will inevitably be revisited through a new bargaining process between monopoly and competition rationales.

2. Universal service: a paradigm rather than a strict definition

Talking about universal service is firstly a matter of entering a semantic field. Discussion of the definition of the concept will at some time or other make reference to closely or more distantly related concepts of:

- public service
- public network
- public infrastructure
- service of general economic interest
- reserved service
- common carrier
- basic service
- basic public service

In the words of Eli Noam [2], who broaches the problem of ambiguity by reference to the common-carrier concept: "a common carrier need not be a 'public utility' or a 'regulated monopoly' or vice versa ... 'universal-service obligation' is the requirement of a carrier to reach every willing user and desired destination, while common carrier refers to a service obligation given a physical plant.... 'affordable rates' are a monopoly and utility issue; where common carrier is concerned with price it is with relative price levels, to prevent price discrimination".

Although one could thus separate the conceptual spheres in a formal manner, I will argue that this global paradigm usefully underlies the debates. Noam recognises this when he grants that: 'in seeking an applicable common-law definition of common carrier, a good deal of confusion results from the long and complicated history of that concept'.

As is often the case when there is no consensus on the definition of concepts or terms, the actors have a different or even conflicting perception of the stakes involved.

To obtain an understanding of these stakes, we shall successively approach the three key concepts – universal service, public service, common carrier – by looking at some recent analyses and then drawing the first lessons from this formal approach.

2.1. Where a 'universal' definition is still missing

Surprising for some, significant for others, there is not an explicit definition of universal service in either the US Communications Act of 1934 or, fifty years on, in the Commission's *Green Paper* of 1987.

Milton Mueller [3] remarks that: "The words 'universal service' never appear in the Communications Act of 1934... at a time when the telephone was viewed as a social necessity".

Jill Hills [4] notes clearly that if universal service is not a term used in the 1934 Communications Act, the Act introduces however the concept of both quality and affordability together with a geographical basis of provision.

Today, it is generally accepted that universal service implies that a ubiquitous communications infrastructure can contribute to national unity and equality of opportunity.

Thus, the common notion of universal service, which links high household penetration with the separation and settlement practices of a regulated telephone monopoly, seems to be a very recent construction.

Very recently, the CEC, pressed by the European public-network operators, attempted to define the content of an updated universal service. In the Communication of 28 April 1993, the political goal of universal service was identified as: 'making available a defined minimum service of specified quality to all users at an affordable price'. The Council Resolution recognised that 'the application of ONP (open-network provision) measures constitutes the basis for a definition of universal service'. On the basis of ONP, the universal-service concept includes:

- the basic provision of service
- quality of service
- tariff principle: cost-based plus flexibility
- dispute-resolution mechanisms
- special public-service features (emergency)
- certain Union-wide service features

But this approach defines more a regulatory framework than a precise content and a concept delineation.

2.1.1. *Universal service is more closely linked to the public-service concept*

The recent White Paper, named Rapport Dandelot [5], on the French telecommunications policy for the nineties, calls universal service a subset of public-service obligations: 'the grand principles of equality and continuity of service are established by our law. They otherwise comprise nothing specific to France and are applicable in comparable forms to public-telecommunications operators all over the world. Moreover, traditional public operators have a universality obligation which implies offering their services to the entire population'.

It must nonetheless be recognised that 'the great difficulty inherent in telecommunications is that the scope of public-service missions remains imprecise: the term is employed everywhere but its substance has not been defined'.

Marc Dandelot reminds us that the universality of the telephone service has been won at the expense of a very artificial pricing structure. Other public services do not practise this type of cross-subsidy.

2.1.2. *The common-carrier concept, overlapping the universal-service concept*

The common-carrier concept was born in Europe with a broad meaning of 'open to public service'. In moving to the United States, the definition, scope and principles gained in precision.

In an article entitled: 'System integration and the impending doom of common carriage' [2], E. Noam gives us the key elements for understanding the stakes related to common-carriage deployment.

The scope of common-carrier activity can be defined as the following five items:

- service is regular
- customer demands are not readily predictable and are changeable
- the carrier solicits business from the general public
- it is concerned with relative price levels to prevent price discrimination
- law and regulation define the responsibilities of the parties

In a certain manner, these criteria could constitute a subset of those applied by the European Commission in defining universal service.

According to Noam, the prohibition of unreasonable discrimination is the most important component of the common-carrier obligation. By definition, the common carrier offers service on a non-discriminatory basis, neutral as to use and user. It is an arrangement that promotes interconnection, encourages competition and reduces transaction costs.

2.1.3. *Where overlapping is meaningful*

It is obvious that in the light of previous developments the notions of universal service, public service and common carrier are inter-related.

If each one could be thought to determine a reference plane, it must be recognised that hitherto, the three planes have been more or less superimposed, with the possibility of the upper plane being one of the three notions, according to circumstances, the political context, national culture or economic demands.

This is why to my view, a static definition should not be retained. To define universal service it is necessary to measure the degree of interference with the related concepts of public service and common carrier. This interference is often at the root of competition policies. If not properly mastered it could maintain a certain amount of confusion in the debate on and ineffectiveness of public policies.

2.2. *Competition is key to the rationale of universal-service evolution*

Competition has and will play a decisive role – and this is only outwardly paradoxical – in the development of the very concept of universal service.

Hills stresses that nowadays the concept is under threat from privatisation and the introduction of competition into the network, all of which produce incentives for PTTs to alter their corporate strategy. Dandelot recently noted that it is competition that compels us to redefine a concept hitherto containing very little substance.

Noam remarks that 'the emergence of a partly competing, partly collaborating network of networks is a foundation for a change that is fundamental in the long term' and that, in such a context, the erosion of common carrier is much more likely happen.

2.2.1. *The Theodore Vail vision*

Mueller puts into historical perspective the relationship between competition and universal service. He reminds us that the words 'universal service' entered the vocabulary of the United States telephone regulation in 1907, when a competitive struggle between the Bell system and independent companies was at its peak. The universality of the telephone network became an issue at that time because the competing systems were not interconnected with each other.

Access competition promoted a universal telephone infrastructure by placing a premium on a network's scope. Had the competitors been interconnected, the incentives to pursue universality would have been greatly weakened.

Vail conceived universal service as an integrated monopoly that could interconnect all telephone users – the service should be provided under control of a single firm – and that monopoly was the best way to achieve it. 'Universal service became a competitive strategy, a political slogan and a catchy advertising term all in one.' More recently, the redefinition of universal service occurred when cross-subsidy practices were threatened by the development of competition in the 1970s and 1980s.

In a liberalisation-policy context, the imposition of high rental charges, first encouraged by the OECD and thereafter by the Commission, on the basis of the so-called 'cost-based tariffing principle', was the indirect trigger to reconsider universal-service obligations.

In a parallel development, the European Court of Justice has recently held that the maintenance of exclusive rights is only justified to the extent that the resulting restrictions on competition are necessary to ensure the performance of a task of general economic interest which has been entrusted to the undertaking enjoying those exclusive rights.

This could perfectly well lead to a strengthening of the preserve (that of public service): while demanding a comparable level of universal-service obligations throughout the Union, the Commission accepts the possibility of member states imposing complementary obligations on the basis of their specific social objectives, as long as such obligations do not create barriers to market entry.

The maintenance of exclusive rights and the absolute need to envisage the interoperability conditions to be applied to infrastructures, and not necessarily to services, could result in a scenario in which monopolistic competition becomes the dominant form of organisation of the markets for telecommunications services including universal service.

2.2.2. *The essential techno-economic dimension*

The success of the universal-service concept was due to the support for its strong techno-economic dimension.

As Mueller points out, the 'one system, one policy, universal service', as Vail required, meant a nationally-interconnected, centrally-coordinated monopoly like Western Union. Far from pursuing social ubiquity in the modern sense, the Bell system was modelling itself on the telegraph system.

The implementation of mechanical and electromechanical switching enabled universal service to show its full force. The strength of the concept lies in a highly consistent techno-economic system that can be summarised as follows: *hierarchical network-economies of scale-positive externalities.*

At the heart of the universal-service debate, present thinking on the cross-subsidy of tariffs is still very much tied to the public-network structure inherited from the first half of the 20th century: local, trunk, international. In the long term, however, this architecture no longer has the same relevance:

- technology makes the cost of using infrastructures independent of distance;
- in addition to reroutage possibilities, software affords far more revolutionary possibilities for a fractional – even fractal – supply.

The introduction and management of economies of scope had already seriously dented the model; under the combined effect of resale, interconnection and competition, the common-carrier status became almost 'unsustainable' from the beginning of the 1980s.

The paradigm of universal service, as envisaged by Vail, introduced a form of stability in technology and especially in network engineering (the switching-transmission couple). Today, this is no longer the case: the unpredictable technology trend makes it difficult to have a clear vision of the relationship between infrastructure, networks and services.

Consequently, and contrary to that which prevailed when the first infrastructures were established, as a support for universal service, economy of access becomes quite different to economy of usage.

Attempting to optimise a production function which integrates the two, now constitutes an exercise full of peril. An attempt to continue to regulate the system with unsuitable tools for this new reality, runs the risk of proving harmful in terms of economic efficiency and, beyond that, prejudicial to the defence of public interest.

The Council Resolution which recognised that 'the application of ONP measures constitutes the basis for definition of universal service' conveys the actual difficulty of challenging the validity of a regulatory approach which carries the seeds of some degree of inefficiency.

2.3. *Universal service: a bargaining process*

It is to be noted here that, while incurring the risk of skimming, competition gives bargaining power to the incumbent with an important policy either to maintain a form of status quo or to exploit his advantage, somewhat like Theodore Vail did in his time. Hills points out, for example, that the support of some regional Bell operating companies (RBOCs) for an extension of basic service was part of a strategy to place pressure on Judge Green for information services, and also a means to gain the support of the Public Utilities Commission for investments they would not otherwise allow.

In the European context, having failed to obtain this status quo, public-network operators have clearly brought up the problem of the 'burden' which universal-service obligations will now constitute. The Commission thus recognised that the cost of universal-service obligations can be shared amongst market participants and that a clearer definition of the universal-service principle is vital in this process.

Along this line, the concept of common carrier appears to have developed as a sort of social compromise whereby a carrier was made to bear a special burden of care in exchange for the privilege of soliciting business from the public.

Noam claims it has permitted society, particularly the US, to entrust its vital highways of information to profit-oriented companies; it was an element in establishing a free flow of information, neutral as to its content; and it reduced the administrative cost and the burden of liability of the carrier.

From a more European viewpoint, common-carriage or public-service obligations may be seen somehow as a 'bargaining' notion, whereby the state tries to manage the public interest in an economic context dominated by the private ownership of transport networks.

2.3.1. *The socio-technical deal*

It would be incomplete and detrimental to the analysis not to look into the universal-service problem and consider only the techno-economic aspect.

Admittedly, in a situation where competition is an essential element in regulating supply, optimising a universal-service production function remains a thorny problem. But the demand function, which we have not yet discussed, stands a chance of becoming just as determining a factor so far as both public policies and operators' behaviour are concerned. One of the main lines of the present debate is, however, the reconceptualisation of universal service in terms of content and consumer rights.

As Hills pointed out, if universal service has traditionally been constructed as a matter of penetration, recently the debate has moved from interconnectivity towards content: what quality of services and what services are to be included within the basic access.

Once the availability of a simple telephone service has been secured, what demand will there be in the nature of universal service? Today, we do not have sufficient resources to answer this question. There will no doubt have to be a phase of social cobbling before the answer emerges clearly. It is the supply of services at a more or less accelerated rate which should allow this demand to show itself. Already, two main lines of thinking could be suggested in regard to the renewal of universal service: the integration of mobility and image. Universal service based exclusively on point-to-point vocal communication seems destined to belong to the past.

2.3.2. *The maintenance of universal service: a headache for regulators*

Interfaced between social preoccupations presented as legitimate or non-legitimate, according to the political powers, and the economic demands which turn the wheels of industry, regulation is seeking new inspiration in terms of doctrine. As Noam points out: 'Opening competition will prove to have been the easy part. Dealing with the consequences will be part of the next and more difficult agenda'.

The real key to the construction of universal service was that it linked a political goal – liberal egalitarianism – to a particular system of economic organisation. Both of them need to be reviewed in the light of the dynamics of the telecommunications industry and what could be called 'the electronic democracy'.

Certain arrangements that have been suggested, for example by the Commission, may not stand the test of reality. New ideas and suggestions need to be promoted whenever they are based on serious analyses. Some will emanate from a deductive analysis. This applies for instance to a proposal made by Noam in conclusion of his analysis of the common carrier/contract carrier dual system: 'The nature of telecom networks permits dual operations on each end of the transmission path ... the upstream end could be contractual, while downstream users would be common carriage'.

Another way of gaining accuracy in this domain is to scrutinise key concrete developments in the telecommunications arena, which present elements for reconsidering the universal-service perspective. This is the objective of the inductive second part of this article.

3. Beyond the POTS: an industry perspective

It must be emphasised that the renewal of the contours and content ascribed to the universal-service concept involves a renewal of our analysis grids. This is what we attempted to introduce in the first part, which was devoted to the formal debate. In this second part, we leave the politico-academic debate so as to examine four symbolic cases taken from the recent industry evolution.

We have chosen to draw lessons from four industrial experiments which nevertheless are still in progress. Each could in its own way provide us with useful information on what the universal service of tomorrow could or could not be. The cases selected are ISDN, Internet, Iridium and PCN/PCS. They are indeed four technical communications systems which, in one way or another, at one time or another, have run as candidates for universal service.

It is exactly this 'in one way or another, at one time or another' which is going to be of interest to us. We shall present some thoughts case by case while keeping to some chronological order which, in itself, is quite revealing as to the manner in which the universal-service problem recurs in the field.

3.1. The provisional but foreseeable failure of ISDN, as a universal service

Called upon to regroup under the Comité Consultatif International de Télégraphique et de Téléphonique (CCITT) banner, the PTTs proclaimed, for early ISDN, their allegiance to a standard from which the network and service strategy would be easily derived. In practice, this theoretical view came up against a certain number of difficulties that reveal a lot about the rationale of the telecommunications industry [6].

3.1.1. The techno-economic challenge

Due to the emphasis being put on digitalisation, network operators must answer a question with great strategic implications: which definition should be retained for the nature of the ISDN network?

The first option consists of making a new network (identified as a particular product), a new value-added network, placed side by side with other networks managed by the operator.

The second option consists of making a meta-network, having overall responsibility for and linking all the different networks, switched networks, packet networks, leased lines.

The last option consists of making the telecommunications network, ie, the outcome of a universal-network logic and the substitute for all the dedicated networks.

The choice of one of these three options is closely linked to the operator's culture, and objective technical development (in particular the speed of digitalisation that is mentioned above). International standardisation, from the subscriber terminal to the switching, passing through transmission, signals and security, enables ISDN to be treated homogeneously, and leaves open every opportunity to place oneself on one of these three dimensions: value-added network, meta-network or universal service.

Thus, British Telecom seems to have opted for the first definition, Deutsche Telekom wishes to position ISDN on a meta-network dimension, and France Télécom has chosen the universal-service position to start with.

Until now, market evaluations have been too optimistic. This is due to the method that quantifies the market in terms of ISDN line-penetration rates. This approach, consistent with the universal-service model, that wants to homogenise user expectations and stipulates that ISDN is a substitute for existing services, in fact makes it even more difficult to forecast traffic and limits the interest in market segmentation.

Public operators have shown some incoherence in their entry strategy. As a consequence, there has not been a clear reply to questions such as:

- To what extent does the introduction of ISDN require collaboration between suppliers?
- Up to what point should demand-creation and market-priming be carried out?
- How should standardisation and service-integration be managed?
- How far and fast should tariff policies evolve from the take-off point?

3.1.2. The limitations of the old ISDN paradigm

There should be a strategic repositioning of the main industrial actors and customers as well, once they overcome the fear of committing themselves to (and therefore invest in) a dead-end solution.

The calling into question of the model envisaged by the CCITT seems to have a central issue of vital importance – the matter of interdependence between the definition of services, entry strategies and demand.

The typology of ISDN services and of their definition has been fixed by international bodies coming under the International Telecommunications Union. Their content in Europe has been specified by a certain number of committees under the aegis of the European Commission – in particular Groupe d'Analyse et de Prévision (analysis and forecasting group) (GAP) – or of the Confederation of European Posts and Telecommunications (CEPT).

These bodies, where technical culture is dominant, have at no time known how to measure the impact that the introduction of competition, leading to an indirect challenging of the very foundations of the universal-service concept, would have on the deployment of ISDN. It was a long-held belief that the standardisation process which was started at the beginning of the 1980s would be sufficient to establish ISDN as the natural heir to public switched telephone networks as a universal-service basis.

In fact, the CCITT approach to services (which was established for ISDN purposes and is still alive in research on broadband technologies) is of limited value in relation to user needs. This is partly a function of the distinction between bearer and tele-services which are purely technical, rather than user or business concepts.

In some respects, the Euro-ISDN policy, which the Commission is implementing with the more or less active support of European operators, is following the same path even though it possesses certain virtues.

However, in breaking up the interdependence of the universal-service, public-service and common-carrier concepts, deregulation has created a kind of point of no return even though it has not launched an actual frontal attack on universal service.

Privatisation of the public operator has thus caused the latter to change views in which ISDN as a universal service had hardly any appeal. Exposing the public operator to competition in the value-added services segment has spoilt the possibility of standing the ISDN universal service against the common-carrier principle as is shown by the conflict bound up with access to channel D by non-discriminated users.

ISDN – which is potentially the last network to have been set up by a highly concentrated PTT structure – has found real difficulties in its promotion phase, and a resulting slow level of take-up, possibly partly due to a lack of understanding about how user services should be defined and provided.

The ISDN scenario, as a reference to universal access and service, is now caught between the development of corporate networks interconnection, based on closed user-group communication patterns and requirements, and the deployment of videocommunications, closer to the mass media rationale.

3.2. *Internet: an inconsistent approach to the techno-economic dimension*

Based on the NSF (National Science Foundation) network created in 1988, Internet now embraces 4,976 networks interconnected on the one hand according to the TCP/IP (transmission-control protocol/Internet protocol), and on the other by access gateways in the case of proprietary systems [7].

The Internet network has achieved an international dimension which continues to grow simply due to the connection of new networks (TCP/IP non-proprietary).

While at first only researchers were authorised to connect up to this network, firms and even individuals are now able to connect their computers to Internet. All the same, this connection can be made only if the access charters are respected (NSFnet charter, EUnet charter).

Internet has thus in a very short space of time changed from a closed-user logic to a universality logic, and this has been done under the simple pressure of what can be called a social demand.

On its techno-economic foundation, the 'Internet System' – PSTN (public switched telephone network), gateways, hosts, backbone – functions on the universal-service pattern (a universal service of data and not voice communication) but it does not have the status and is not at the moment recognised as such

The great majority of Internet users obtain access to mail and file-transfer services via the switched-telephone network.

However, unlike a public service, the connection is paid for by the user. Internet does not come to the user – it is the user who establishes his link to obtain access to Internet (by telephone line, packet-switched data network or leased line).

Even so, in the case of Internet most of the ingredients can be found which define universal service, including what it imagines as cross-checking with the common-carrier and public-service concepts.

It is this proximity which could have made it possible to accept the proposal of the Electronic Frontier Foundation (EFF), close to Apple, to make Internet a national public network.

It is also this cousinship which permits Internet to run for election as a support network for the 'electronic highway' which is so cherished by the present American administration.

What I believe will prevent Internet from playing this universal-service rationale right through to the end, and thus to its advantage, is that it belongs in an open environment which bridles the application of Theodore Vail's doctrine, where it would have the exclusivity of control, a more restrictive but at the same time more decisive element. In concrete terms, this means:

- An ambitious interconnection policy but one which makes the system more vulnerable to developments within its sphere and, in particular, that of its competitors.

  This policy has initially taken the form of a demand for interconnection of private firms – 'IP means business'. This calls for a very rapid increase in capacity to reach 45 Mbit/s in 1992 and 155 Mbit/s in 1994. The implementation date for this policy will also be chosen in the light of current developments within the public networks.

  In December 1992, the IETF (Internet Engineering Task Force) announced that within the month it was going to publish a standard permitting the communication of data between TCP/IP and asynchronous transfer mode (ATM) networks, the Internet-user protocol standard for ATM. This standard was designed exclusively for the communication of data, although the IETF had the intention of adapting it to vocal and audiovisual communication to permit multimedia applications.
- An opening of services which will induce Internet to adopt a value-added network logic whose techno-economic paradigm corresponds less and less to that on which universal service has been developed. Conceived originally as a network designed exclusively for research, the NSFnet now allows for commercial services.

  On the supply side, efforts are quite naturally given an international orientation. Among the main international commercial IP initiators are the following:
  - Sprint, the first operator to exploit a commercial IP network, who in September 1992 envisaged internationalising their services through an X.25 access to their network;
  - Unipalm Ltd (GB) and Swipnet (Sweden), two pioneering providers of commercial services on IP networks who, in September 1992, contemplated establishing a pan-European commercial IP service;
  - BT North America Inc. who planned to offer dial-up asynchronous access to IP hosts from its nearly 5,000 nodes worldwide by the end of 1992;
  - Infonet Services Corp's Infoland IP service who envisaged serving eleven European cities by the end of 1992;
  - Global Enterprise Services Inc. who formed a subsidiary at Princeton University to provide commercial IP services.

  Despite the creation of a business firm with the aim of endowing services with greater legibility, Internet bears the seeds of a transformation hardly compatible with the traditional universal-service model which, for a lack of stability in the available technical equipment, necessitates a satisfactory degree of predictability to compensate for the freedom of use which is accorded a priori and guaranteed on principle.

  The creation of a joint venture between IBM, Merit and MCI, ANS (Advanced Network and Service Inc.) indicates the veiled, growing risks of an under-optimisation of the production function: while ANS services have hitherto been limited by special access to Internet, ANS has developed its own lines and is seeking its own connection points in Europe.

3.3. The Iridium project, non-legitimate candidate for universal service

Motorola's Iridium project is sufficiently well known and therefore need not be formally presented.

3.3.1. Strategic positioning

The project started on a 'custom-built' footing, employing the club effect produced by an offer of services which at the outset was intended to be completely independent of that of other operators. (Obviously this did not prevent links with terrestrial infrastructures to connect calls to fixed correspondents, although these interfaces only came into play in the last link of the communication chain.) It was definitely a 'service' role so far as the infrastructure remained de facto completely transparent for the user: what was sold was a service allowing communication with anyone at all, anywhere in the world [8].

This positioning relied on several well-targeted customer segments for making profitable a technical system which, although extremely efficient, was also very costly and could only be marketed at a high price. It was supposed that, strongly motivated by a service which would enable them to solve all their problems of worldwide communications, these segments would have a very weak price elasticity, as did the first subscribers to cellular-radiocommunications systems or to the telephone.

The reasons that led Motorola to develop the role of an Iridium-services offer are numerous. Undoubtedly, however, they are primarily bound up with the fact that a services-offer of this nature by a supranational operator was too early for currently operational industrial models. It is impossible today to go against traditional telecommunications operators: they must be accepted. There was particularly strong opposition on the part of operators from developing countries during the WARC (World Conference on the Allocation of Frequencies) in February 1992. They feared the introduction into such a system of a regular diversion of international communications, which constitute their principal sources of revenue (allowing them in return to guarantee the development of a public-telecommunications service for the entire population).

Opposition also presented itself less openly but equally significantly on the part of operators from developed countries. This was in the form of a marked disinterest in regard to Motorola's numerous approaches within the framework of their canvassing for investors. Although it is true that operators in developed countries have hardly any fears of a system of this nature, which would be demolished on their home grounds by the universally less costly cellular-radio-communications services, they do not appear ready to guarantee the development of such a service offer. It is interesting to note that the duopolistic situations which obtain in most developed countries in regard to the provision of mobile-telephony services have not allowed Motorola to establish itself. Other investments seem to take priority over those in a system with an international calling.

Rather than consider the impact of a competition policy, it would perhaps be better to talk about the effect of regulation models or regulatory mechanisms which have not yet been sufficiently developed to be integrated in an offer of this type. It is in any case quite clear that it is definitely this impossibility or this refusal by established structures to take charge of the emergence of a new service category that has obliged Iridium to move from a relatively revolutionary offer of a 'custom-built' service to a more traditional universal-network offer. It seems to us, however, that these industrial models are bound up with the competition policies implemented in the different countries and, this being so, it can be seen in the case of Iridium how the definition of a competition policy can go as far as to influence the very nature of a service offer, making the cursor move in the direction of more or less extensive 'universality'.

This change of position by Iridium, however, is equally interesting in that it brings out a contradiction between universal service and public service. The reason why the main obstruction encountered so far has come from operators in developing countries, is that the project runs counter to the development of public service. Iridium can in some way be perceived as an offer of services intended to be universal without having to bear the constraints of a public service. Because operators in developing countries have to finance the development of their own telephone network, they cannot afford to have their international communications bypassed by the action of an operator who would not contribute to the development of local infrastructures.

Thus, the reason why Iridium has no place today is two-fold: universality changes its scale with a project of this type; and this change of scale in some way puts the service in competition with the traditional telephone service (whether or not including terrestrial mobile services, according to country), which either remains to be developed or constitutes the chief source of revenue for established operators.

3.3.2. Issues to be considered

This could lead us to raise the following questions:

- Considering the regulatory model now in the process of being set up, is the mixture of a 'custom-built' offer in terms of clientele, and a universal offer in terms of geographic coverage a viable one? In particular, can it be provided by a single operator in a world context which is still largely monopolistic or duopolistic? The question deserves to be raised: should there not be some examination of the possibility of breaking up public service or universal service into a certain number of distinct sub-elements open to reconfiguration in an offer of a non-public nature, and one degree less universal, but likely to be proposed by operators other than traditional ones?
- At the same time, is it conceivable that this recomposition could be carried out? Do not the constraints which weigh on the offer of public services prevent de facto the emergence of such an offer, actually necessitating alliances with the operators currently responsible for public service? Must our thinking in this case be changed so as to remove these constraints, or so as to integrate them in a universal-service offer in a manner different from believing that one single operator should take charge of these two offers? In this case, is the notion of access charge, which permits service operators to contribute to the financing of the distribution infrastructure to which they have access, being given sufficient thought?

It can be seen that to question oneself on the principles for defining the concepts of public and universal service and on the possible impact of the organisation of markets (and thus implicitly, that of competition) on these definitions on the basis of an 'outsider' offer of services in relation to the traditional offer by terrestrial operators, does not fail to raise certain other questions. Some of these could appear, a priori, somewhat fanciful, but there are others which confirm the thought that has been given over the past few years to the notion of public service which represents a crucial question today for both regulators and public-service operators.

3.4. PCN/PCS : where universal-service rationale is facing multiple-network arrangements

The case of PCN represents a good complement to the preceding cases, as it reveals a quite symbolic problem of certain industrial dynamics at work in the telecommunications field. I shall sum up this problem concisely: competition between technologies for unspecified services. This is more or less contrary to that which until recently was the foundation of universal service.

This unsolved equation requires us to seek a collective optimum, given a priori, in the economy of networks and market structures [9].

With regard to the organisation of markets for personal-communications services, there are in fact two possible options:

- the first consists of creating one or more new radio networks (high degree of autonomy in terms of infrastructure, network logic);
- the second consists of favouring radio access to fixed networks (PSTN, cable networks) and existing (service logic) mobiles (analog, group special mobile (GSM)).

3.4.1. *The creation of new radio networks*

Pragmatically, this option is seen in Europe in the granting of DCS 1800 authorisation to operators other than GSM operators. Several providers or a liberalised market is perceived here as the best means of guaranteeing the development of the mobile-communications market. In this context, DCS 1800 operators are compelled to accept an obligation to provide eventually a national coverage, and are supposed to offer a universal service (coverage + cost) in a very short time. However, the uncertainties over this are becoming increasingly clear:

- penetration of the general-user market by personal-communications systems is bound up with an overall utilisation cost close to that of the ordinary telephone...
- ... so DCS 1800 operators must quickly build a large market so that the effect of size will allow them to reduce the cost of terminals to a point that puts the service on competing terms with the ordinary telephone service.

It is at this level that we can see the limits of such a scenario of development in the direction of a universal-communications service. A service of the PCN/DCS 1800 type offers more or less the same functions as the GSM and consequently the initial costs of GSM and DCS 1800 terminals are very close: this acts as a brake on service penetration. In addition, DCS 1800 networks bear a cost of access to fixed-network resources identical to that of GSM, while the DCS 1800 infrastructure costs are two to four times those of a GSM infrastructure providing similar coverage.

Adding to these problems of cost is the strategy developed by GSM operators to counter the DCS 1800 offer: the micro-cellularisation of GSM networks; and a commercial offer (described as 'type MCN' (a Vodafone concept) which consists of offering a service specially designed for private subscribers at reduced tariffs, compared with the classic GSM offer), for all calls coming from a determined cell. In that it permits GSM operators to reach new segments of the market (universalisation of the GSM service in terms of price and coverage), while reducing the costs bound up with localisation and cell-transfer equipment, this strategy and the effect of its announcement has forcefully reoriented the functioning logic of 'PCN universal' network operators.

3.4.2. *Promoting radio access to existing mobile and fixed networks*

The emergence of PCS services in accordance with this option, covers three principal choices: connecting wireless gateposts to fixed networks (CT2 (cordless telephone)/DECT) and mobiles (DECT); considering the DCS 1800 as a complement to GSM in terms of both infrastructure and sales offer; and consistently utilising the entire range of existing networks (for example, the American example).

3.4.2.1. *Connection of wireless gateposts*

A CT2 or DECT service could be utilised either as a wireless PABX for professional users (custom-built service logic) or for use in the street in Telepoint applications (general public universal-service logic in the manner of the French Bi-Bop).

The French situation provides an excellent example of the conflict that could occur between a 'public service' and a 'universal-service candidate'. The current Pointel schedule of conditions allows an operator to install CT2 posts that can be connected to the wire network or to France Télécom's Pointel network. It is the conditions of interconnection to the wire network or the Pointel network which determine the economy of a CT2 service, whereas the present Bi-Bop tariffs do not guarantee adequate profitability for an outside operator. Moreover, the possibility of enriching the CT2 network with other services, such as one-way paging, is at present available only to the public operator owing to current authorisations. Finally, present conditions do not allow a new CT2 operator to compete on equal terms with France Télécom's Bi-Bop service. In fact, this 'public service' of universal calling finds itself protected from competition, which could provide an extra stimulus for a rapid universalisation of a service of this type. In the light of the failure in Britain of Telepoint-type services (notably because of an excess of operators), one might well wonder whether the French regulator was not attempting to protect a 'public service' by offering it real chances for attaining universality in the medium term.

3.4.2.2. The PCN/DCS 1800 as a complement to the GSM

In this choice, the GSM networks serve as support networks for numerous other networks and additional mobile services, each one serving specific segments of the population. Taking into account their heavy involvement in the cellular-radiocommunications sector and the high level of investments, European GSM operators considered the introduction of the PCN as a gradual extension of GSM-service applications (product diversification) vis-a-vis their existing clientele, or in the direction of specific uses and target markets, such as the mass professional market and certain segments of the general public. On the technical plane, this evolutionary strategy translates into a reconfiguration of GSM infrastructures to an offer of a PCN-like service (micro-cellularisation in urban zones modelled on the technical strategy adopted by Vodafone with the MCN) and into a territorial-coverage strategy that varies widely, depending on the appreciation by existing operators of the market opportunities and the elements of the competition game.

3.4.2.3. Systematic infrastructure sharing: what savings of means?

The United States is the country in which the most intense thought has been given up to now to the use of existing infrastructures for providing PCS services. Moreover, the FCC recently developed a model aimed at assessing the simulated costs of the installation and exploitation of a dedicated PCS network and the costs of exploiting PCS services using existing infrastructures such as the PSTN, cable networks and radiotelephone networks. For a long time, local telephone companies, cable-network operators and cellular operators have all been maintaining that their networks are in the best position to offer PCS services. One of the principal conclusions of the FCC study is the presence of significant economies of scale between PCS services and the services offered on the three networks mentioned above. According to the FCC, the existence of these economies of scale would, to start with, justify the allocation of PCS licences (and thus of part of the associated spectrum) to radiotelephone operators, to the BOCs and local exchange companies, provided that there is a sufficiently competitive environment.

Independent providers of PCS services will thus be induced to negotiate the use of existing networks, and with that come the delicate problems of organisation of the competition. Telephone and cellular operators may consider that PCS services compete with their provision of radio and wire telephony and that they could reduce the quality of the interconnection they offer PCS-service providers. (One condition for the success of PCS services is the interconnection to telephone and cellular networks in optimum technical and economic conditions.) The core of the problem is thus to decide whether local operators will or will not be confined to the role of carrier's carrier for other PCS providers. It is obvious that telephone operators will all the more rapidly adapt their networks to PCS-service provision so that they will themselves be in a position to offer these services directly. The same type of problem applies to independent cellular operators and, to a lesser extent, to cable operators.

3.4.3. *The economic viability of the project*

In the face of the constraints of economic viability of universal personal-communications services, two major approaches thus stand out, the one not excluding the other:

- the first consists of providing these services by relying on networks already installed, which will make it possible to meet the constraints of accessibility in regard to both coverage (the PSTN and the GSM will have national coverage) and price (the upgrading of existing resources reduces initial investment costs and thus the price of the service);
- the second consists of conferring 'public-service' status on this type of service so as to be able to justify a monopolistic-type organisation (the Bi-Bop service benefits more or less from such protection). This will permit the public operator to implement an ambitious policy of national coverage and affordable prices for the general public (the Bi-Bop service benefits from very advantageous conditions of interconnection with the PSTN which are probably bound up with the similarity between the operator of the PSTN and the Bi-Bop).

In the light of the preceding developments, it can be noted that at present there are no universal systems capable of meeting the needs of personal communications. Competition between operators or distribution networks introduces an extra dimension which leads us to consider the personal-communications universal service as an aggregate of several services associating several multiple networks exploited by different operators.

It is unlikely that this type of approach would be compatible with the requirements that have hitherto been associated with a universal-service offer. The case of PCS invites us to reconsider the techno-economic coherence that lies at the heart of the universal-service concept.

4. Where a new vision for the concept of universal service is required

The success of the universal-service concept in these past fifty years lay in the coherence between its techno-economic foundation and the way it was embedded in the competitive environment, indeed a monopolistic market structure.

Let us recall that this success was based on the optimisation of the production function, where cross-subsidy practices play and important but non-determinant role:

- Network engineering, under the aegis of a public or quasi-public monopoly – the common carrier – took its part in this success story.
- The steadiness of technologies, and particularly crossbar and electro-mechanical switching, made it possible with engineering know-how, to have a full deployment of a national interlinked network.
- Lastly, centralised control of the production process was key to the adequate performance of universal service.

It is also useful, for a complete understanding, to remind ourselves that in a context of a supply which avoids any discrimination against the customer (or the access), and the user (or the usage), controlling the demand process was not a major issue.

In the first part of this article we tried to point out the strength of the universal-service paradigm: techno-economic consistence, competitive strategy, legitimacy of control – even when bargaining with policy makers. Today, those three pillars of the system are becoming fragile with the dynamic of the industry.

So far, no candidate system has succeeded in gaining the universal-service status. The case studies in section 3 have illustrated where the problem lies:

- in the techno-economic dimension, for PCN/PCS
- in the inclusion in the competitive environment, for ISDN
- in the legitimacy of control, for Iridium

The former paradigm – monopoly of service/oligopoly for equipment – which was the foundation of the model implicit in regulation up to the end of the seventies, has had to make room for a new paradigm which today is poorly understood, rather unstable and which gives corporate strategies a degree of unpredictability in regard to both their definition and their impact.

In fact, three factors combined at the end of the seventies to destabilise the former system:

- deregulation, which mainly called into question the natural monopoly in the telecommunications service;
- technical evolution, which with the help of the motor provided by digitisation of the signal will, on the basis of a unifying logic, paradoxically have led to an explosion in the supply of technical systems;
- the change in demand behaviour, which appears to have ceased to be the hostage of the technological supply.

These three factors, which are of course interdependent, worked in the same direction. Uncertainty arose as to the rules of the economic game – an uncertainty that was all the more strongly felt as it was linked to a situation that was particularly new to all the actors in the system who were used to playing according to unchanging game rules.

Will universal service survive these changes? If it does, and assuming that the old-concept paradigm will have to evolve, on what basis should it be refounded?

The future universal-service paradigm will not be defined aside from the rationale which is emerging now, and underlies the whole industry evolution. Several issues have to be tackled.

4.1. *Integration at the heart of the network*

Today, there are two approaches to the notion of integration. The first makes reference to a form of integration of media on a single support element – television, press, radio and telephone, all mixed up within the telecommunications network. From certain aspects, this approach could tie up with developments leading to the hypermedia concept.

The second, more technical, bears reference to the integration of communications functions, whether a single function does the 'work' previously carried out by several (here we are thinking of the evolution of the multiplexing function, which is 'biting into' the switching and network-administration functions) or whether the 'work' mode of different functions no longer enables a clear distinction to be made among them.

In an integrated digital network, switching and transmission operate in real time. There is thus, no reason to separate the two functions (particularly from the point of view of the user of such networks). This being so, control becomes a key function, because this function shows itself to be very difficult to manage and thus to optimise. Only one operator 'intimately' linked to the network logic is capable of doing so (at least he can claim to be).

4.2. *Virtual networks and localisation of intelligence*

The virtual network, connectionless, may be considered as a major option for future telecommunications networks. If switching becomes far more dependent on the processing capacities of user terminals than it is today, and the relative dependence on network storage capacities therefore decreases, the intelligence will then have to glide towards the subscriber network with the general network becoming more economic in usage and easier to maintain. The client will be able to programme the network so as to overcome the problems of access, while the network will furnish functioning reports in real time and in a dynamic manner.

Obviously, the localisation of intelligence remains a major stake in an approach of this kind: will it be so in the actual applications, in the network, in the functions or in the equipment? Control of this scenario will be put in the hands of different actors according to the reply given to this question.

In speaking today about virtual networks, it is in the sense where the same function can be activated either by the public-network operator or by the client within his private network.

4.3. *The frontiers between private and public networks*

The concept of media shared among a certain number of users in the public sector, which was introduced by broadband communications, upsets the customary rules of public-network organisation. Obviously, this causes certain problems which remain to be clarified at a European level. The discussions being held within the European Telecommunications Standards Institute (Committee NA5-MAN) group provide a good illustration of this complex problem, for the subjects being discussed are:

- Confidentiality of access: this is a primordial function. If effective, it can bring about serious distribution constraints. It must be ensured that in every case, the flow of information is out of the reach of any malevolent user.
- Limit of liability: the norm now being worked out presents certain MAN DQDB (metropolitan area network, distributed queue dual bus) configurations straddling the private and public sectors. It will be necessary to determine the liability of the two parties in case of fault. It will also be necessary to be able to identify the responsible party with certainty in any case of malfunction.
- Evolution of services: it must be ensured that the division between software programmes is sufficiently distinct to allow them to develop independently without either party being penalised.
- Quality of service: inasmuch as the entire structure is not subordinate to the public operator, the problem of quality of service from beginning to end is going to come up. Also the problem concerning the validation of the continuity and maintenance of the public-operator service at the subscriber-network interface remains to be defined.

Even the more long-term approaches to the deployment of broadband technologies implicitly render obsolete the frontier between the public and private sectors.

A useful illustration of this could be obtained from the strategy which NTT in Japan seems to wish to establish. The definition of VI&P-type services is supported by a new technical concept that the Japanese will probably try to promote, the ‘virtual-path network’. It has already been affirmed that the virtual-path network will be shared among private networks and the public switched broadband network.

4.4. *New requirements for the standardisation process*

In a context such as this, it seems justifiable to expect standardisation to be concentrated more on the production process (ie, mainly protocols and functions) than on the products.
All the ‘constraints’ to which it seems that standardisation policies will then necessarily be subjected are in line with a certain attempt to optimise the flexibility of the systems.

Among the paths to follow, the following have been suggested:

- Variability of temporal and spatial resolution: signal reception must allow for a simple, immediate transcoding between the different restitution systems and a wide variety of entry systems with optimisation of processing and storage in the terminal areas of the system.
- Handling of information: the representation of information should permit the invention of new information-capture technologies, pointers towards the user and copyright information, copy-protection codes, non-linear correction factors, algorithms of compression and transformation, etc.
- Interconnection: the format should allow for the production of all types of information and its transmission in networked information-processing systems, transport in switched optical or coaxial-cable systems, in distribution or interactive forms.
- Margin of error and controlled degradation: the format should make it possible to anticipate imperfection in the channels and to allocate differentiated codes of error correction to more or less significant signal components.

Although it would certainly be considered unrefined by the technical specialists in the telecommunications of the future, this typology, in our opinion, well describes the contours that future standardisation policies could assume. It also indicates a possible task-sharing between standardisation (guaranteeing system flexibility) and regulation (administrating the opening of these arrangements).

4.5. *The problems facing regulation policies*

These challenges concern putting the open-environment concept into regulatory practice. They are already related to certain identified problems:

- Tariffing and accounting: the economic and social logic governing the fixing of tariffs, the methods of separate accounting, investment and depreciation policies, for national or corporate networks serving private or national interests, are called into question when these networks carry services that perform a function, multiplexed, sometimes indissociable, sometimes dissociable.
- Legal mechanisms: open networks give rise to new legal problems such as the nature of the bodies capable of dealing with disputes or even lawsuits, the implications with regard to antitrust legislation, the criteria laid down on performance, intellectual-property rights, the status of protocols and standards relating to interfaces, access to clients' private information and to data bases, and the protection of civil rights.
- Implementation and performance: the migration towards an open environment, calls for the setting up of acceptable transition mechanisms which take account of the existing means of production, especially mechanisms which allow for non-discriminatory maintenance policies, network-administration operations and the connection to non-open foreign systems.
- Security: by their nature, open systems will change the degree of vulnerability of networks, accidentally or intentionally. New ways of dealing with jamming and interference and of protecting information and private communications will make for flawless systems.

As Dandelot states: 'Technological evolution should lead us to consider the real content of public service with no taboos ... the same service (especially the telephone service) may be provided in accordance with differing technologies: not everybody will necessarily want to pay the price for fibre optics while the possibility exists of installing very economic local networks thanks to radiocommunications'.

Actually, as we indicated in the previous developments, the problem goes beyond the mere technological issue. This is why it looks so difficult to have a clear vision of the future delineation of the universal-service concept. Two rationales are now interworking: the one inherited from the 'golden time' of universal service, and a more recent one which underlies the new industry dynamic. Our paper was just an attempt to clarify these two rationales.

Notes and references

[1] Commission of the European Communities, Developing universal service for telecommunications in a competitive environment. Draft Communication to the Council and the European environment, August 1993.

[2] Eli Noam, System integration and the impending doom of common carriage. Research paper, International Telecommunications Society, June 1993.

[3] Milton Mueller, Universal service in telephone history: a reconstruction, *Telecom Policy*, July 1993.

[4] Jill Hills, Universal service: a social and a technological construct, *Communications and Strategies*, No. 10, July 1993.

[5] Marc Dandelot, Documentation Française, September 1993.

[6] Jacques Arlandis, ISDN in Europe, *European Telecommunications*, December 1993.

[7] Isabelle Ciupa, Le cas Internet. Research paper, Institut de l'Audiovisuel et des Télecommunications en Europe, September 1993.

[8] Valérie Lepeltier, Systèmes LEOs et Service Universel. Research paper, Institut de l'Audiovisuel et des Télecommunications en Europe, September 1993.

[9] Stéphane Vandoit, PCN/PCS, dynamiques industrielles. Research paper, Institut de l'Audiovisuel et des Télecommunications en Europe, September 1993.

Telecommunication: Exploring Competition
H. Williams and M. Borman (Eds.)
IOS Press, 1994

Chapter 7

Regulatory challenges facing data and consumer protection in telecommunications services in the European Union

Herbert Kubicek
Telecommunications Research Group, University of Bremen, Germany

Abstract: The concept of regulation in telecommunications in Europe is usually restricted to economic issues such as tariffs. However, many transactions that have become the subject of telecommunications services, such as electronic mail, teleshopping and telebanking, are also regulated with respect to other issues, for example, consumer and data protection. The national scope for discretion on regulation within the European Union has been limited by the directive on competition of telecommunications services. Only in the limited field of privacy, has a draft directive been submitted, but not yet decided. This paper discusses the political pros and cons for a regulation of data and consumer protection and suggests several dimensions according to which a regulatory system might be designed. A comprehensive annex illustrates the issues to be regulated as well as regulatory approaches already taken by member states.

1. Introduction

The discussion about regulation in telecommunications in Europe is, in its mainstream, dominated by economists and therefore largely restricted to economic issues which can be quantitatively analysed, such as tariffs, market shares, the degree of competition, etc. Most of these economists believe that the needs of users are best met by competition between suppliers of services with regard to prices, and to the quality of the services. Therefore, they argue for deregulation and liberalisation and against any control of suppliers.

But there are other groups with other perspectives and claims. There are the data-protection authorities and civil-rights groups who argue that the privacy of telecommunications activities is threatened by new technological means and organisational decentralisation and, therefore, ask for reregulation. And there are consumer organisations that acknowledge the reduction of tariffs by competition, but still claim that other aspects have to be regulated in favour of the consumer.

All these groups might refer to basic elements of European treaties and conventions. But so far there has been little exchange of views, either between data-protection authorities and consumer organisations or between these and the economists, be they represented by scientists or politicians. Claims for privacy protection refer to the European Data Protection Convention [1]. This convention is based on Art. 8 of the European Convention on Human Rights, which itself also protects the privacy of communications. With reference to this article, the European Court of Human Rights has made several decisions on data protection and on the privacy of mail and telecommunications [2]. The Commission of the European Communities (CEC) Greenbook on Telecommunications stresses the need for data protection in this area [3]. And two important directives on competition in the markets for telecommunications services [4] and on the implementation of open network provision (ONP) [5] mention data protection as one of four basic requirements in relation to which member states may restrict competition. And in the same year, the CEC submitted a package of drafts of directives and resolutions concerning privacy in general [6], and in the field of telecommunications in particular.

Since 1971, consumer protection has been a policy field with its own administrative structure in the CEC. Since then the Commission has initiated several action programmes. The field of subjects is extremely large. So far, questions of telecommunications have come up in connection with distant selling and electronic-funds transfer. The directives on telecommunications mentioned above do not see any need for action or regulation with regard to consumer protection.

There are also large organisational differences between data protection and consumer protection. While consumer organisations are represented on the EU level by the Bureau Europeen des Unions de Consommateurs (BEUC) which receives financial support from the CEC as well as the European Consumer Law Group (ECLG), there is no official representation with financial support for data-protection authorities. The data-protection officers form an International Conference which has a working group on telecommunications and new media, but no office or permanent staff.

In the past, individual member-state political parties, data-protection authorities, consumer organisations, customers themselves, or courts have demanded certain rules and even the revision of technical features or legal regulation. The topics that have been raised this way, vary very much from country to country. For example, in Germany, the regulation of videotext was a hot issue in the early eighties, followed by itemised billing and calling-line identification within integrated services digital networks (ISDN) [7]. In Belgium, itemised billing was never an issue. Instead, there are concerns with credit reporting and distant selling. In Denmark, itemised billing is also not a subject of great concern, but a law has been passed to protect consumers in connection with payment cards. And even when the same topic has been raised in some countries, the solutions developed may be quite different (for example, itemised billing in France and the Netherlands, see annex, section A.1.2.).

Although there is a broad verbal commitment to privacy protection in telecommunications on a general level, there is also effective resistance against corresponding legislation. For example, the revised version of the draft of May 1990 by the CEC for a special directive has not yet passed the Commission, and a recommendation for privacy protection in telecommunications worked out by an expert group of the Council of Europe in June 1992 has not been decided by the Committee of Ministers, but has instead been referred to another committee. The main argument is that there is no need for new regulation because the existing laws already cover the problems, if any. The basic view of this opposition is that telecommunications is just being liberated from the chains of state-controlled monopolies and that at this stage, the dynamics of innovation set free by deregulation must not be hindered or even stopped by new governmental control in the name of privacy. This argument acts as justification for a general rejection of any new regulation of data or consumer protection, or both, without examining the problems in detail, the different options for handling these problems and their respective effects on the acceptance of services and technological innovation. Therefore, this argument is an ideological one and ironically one that hinders reaching the economic objectives.

We have learned that regulation in the area of environmental protection has stimulated technical innovation, led to new markets and as a net effect has brought about positive economic results. Similar impacts are to be expected for data and consumer protection in telecommunications. In Germany, for example, the economic committee of the Federal Council argued that the acceptance of ISDN among other things was so low because the Deutsche Bundespost and German telecommunications manufacturers had not taken the criticism from the data-protection advocates seriously, thus provoking a long-lasting public debate with negative headlines for ISDN.

There are at least three reasons to examine whether existing legislation on data and consumer protection deals with problems that users of telecommunications services might face (see also Council of Europe [8], Ungerer [9]):

(1) Due to technical innovations, new features and services can be offered which, besides their benefits for some or all users, also pose problems with regard to the privacy of communication. Examples are itemised billing or calling-line identification. These problems usually call for a balancing between the rights of the two parties in communication and, in some cases between the service providers and a fourth party as well. The general data-protection laws usually are only concerned with the protection of an individual against a data-processing unit or governmental authority. But, in communications systems, new problems arise out of the relationship between the participants (see, for example, Roβnagel [10], Noam [11]). Existing law only contains the basic objectives, values and requirements, but these have then to be applied to new situations in telecommunications.

(2) As long as there was only one network operator and service provider who was subject to governmental control, including parliamentary supervision, and watched by the media, it was relatively easy to detect cases where privacy of participants was intruded upon. Articles in major newspapers were often sufficient to have politicians modify technical procedures, organisational rules or legislation And even if issues had to be taken to court, the situation of the monopoly and state control required a balancing of rights in the context of public law, ie, the distribution of the rights between the state and the citizens. After deregulation, private carriers or service providers came under private law. Constitutional or public law dealing with the privacy of communications does not automatically apply to these organisations (Rieβ) [12].

(3) Many so-called value-added network services include transactions which have existed for some time in the form of direct personal or written communications, or both, and are as such subject to different regulations with respect to data or consumer protection. For example:

- the privacy of letters is regulated in mail services;
- the prices of products have to be shown clearly in shops and there are special rules for mail-order shopping;
- payment by cheque is governed by rules regarding security and liability.

All these rules, laid down in different laws, refer to the traditional forms of direct personal or written communication. But many of the elements or procedures that are required do not exist in the electronic equivalent. Therefore, the existing regulations do not apply to these electronic services and thus are causing uncertainty for all parties concerned.

For these reasons, national legislators have reacted in one way or the other. However, regulation at the national level is not sufficient for several reasons:

(1) With the emerging single market for goods and services, transborder communications will increase. Letters will be exchanged electronically with partners in other countries, data bases will be logged into or foreign goods ordered, etc.

(2) In order to increase competition between the providers of telecommunications services, the Council of the EU has passed a set of directives which limit the discretion of national legislators with regard to the admission of service providers from other member states.

(3) So far, member states have the right to restrict the free flow of goods and services in connection with matters of public order, security and health (Art. 66 in connection with Art. 55, see Rieβ, p 748) [12]. In addition, according to the European Court of Justice, restrictions have to be accepted in order to fulfil necessary requirements in the general interest, which explicitly includes consumer protection. Issues of data protection in some of the member states have the quality of basic constitutional rights. All this might lead to different strategies by member countries and thereby to a fragmentation of the market for telecommunications services.

(4) As the requirements of data protection may involve certain technical features of terminals or switching units, including their software, the producers of this technology have a huge interest in identical requirements within the whole market because diverging requirements reduce the economies of scale which were to be increased by the EU telecommunications policy.

The last argument is of particular importance for the debate mentioned above on the relationship between the economic and the data protection view: not only a general, but a specific regulation in data protection is necessary in order to realise the single market for telecommunications technology.

So far, these arguments have not been strong enough to overcome the opposition against the regulation of data and consumer protection. Members of DG XIII of the CEC would not promise that the revised draft of the ISDN directive would be passed by the Commission. In the resolution of the Council on necessary future developments in telecommunications of July 1993, data and consumer protection are not even mentioned [13].

Meanwhile, different member states have already taken action. Thus, there is the risk of diverging technical requirements and legal regulation. The pressure at the national level, which leads to regulation there, apparently does not reach the Union level. While all the member states share the economic view of a single telecommunications market, there are big differences between the relative importance given to questions of data and consumer protection. And even if there was consensus on the priorities and the objectives to be reached, the organisational and legal procedures remain so different, that consent on detailed requirements would be extremely unlikely. But other difficulties have already been overcome in the process of restructuring telecommunications. Therefore, one should not give up before a detailed analysis of the problems and the options for solving them has been carried out.

Perhaps the fear of large, new administrations, and a number of new laws in the German tradition is one of the reasons why some member states oppose regulation of data and consumer protection. Therefore, it would seem appropriate to look at other options.

These options refer to the:

- level of regulation (national/Union);
- the legal area (telecommunications or the respective transaction);
- the kind of regulatory institutions and their competence, ie, whether there should be only one organisation for all regulatory matters and whether there should be an evaluation of data and consumer protection issues within a procedure of admission of new services or within a procedure of certification.

A review of the data protection and associated problems related to telephone service, network operators and value-added networks is set out in the annex to this paper. This annex considers the different issues raised by member states and details some of the regulatory measures that have already been taken. This review also identifies the need for harmonisation of regulations at an EU level as well as the difficulties any such effort would face.

2. The hard way to an appropriate regulatory framework

2.1. The basic regulatory dilemma

The examples given in the annex make it clear that Union-wide regulation is necessary in order to promote a common market for telecommunications services, for at least three good reasons.

(1) Users of the services mentioned in Europe can claim the protection of their fundamental personal rights according to the Declaration of Human Rights by the Council of Europe, which includes the privacy of communications, and according to the data protection

declaration. As services are offered across national borders, this European level of regulation becomes more important. And the challenge is already being taken up by the Council of Europe and by some units within the CEC.

(2) The telecommunications policy of the EU will offer service providers and manufacturers of equipment the possibility of one big market, allowing for economies of scale which will lead to a stronger competitive position in international markets. This means that different technical requirements are not permitted in the member states, because national adaptation raises the costs of development and production. Instead, the technical requirements must be identical, and they must be predictable in order to realise the goals of industrial policy in this area.

(3) As long as there were monopolies and the number of actors and services was rather low, informed agreements between PTTs or governments, or both, were appropriate in most cases. However, with deregulation the number of services and actors is exploding. Deregulation has created the need for new regulation.

However, it has also been shown that there are at least three factors which make it extremely difficult, if not unlikely, that a common regulation of data and consumer protection will come about, which will lead to calculable and specific requirements for all present and future telecommunications services and equipment:

(1) The relevant values are quite different in the member states, in particular with regard to privacy or data protection aspects of institutions and procedures.

(2) The legal systems are very different: in general with regard to institutions and procedures; and in particular with respect to telecommunications and data and consumer protection.

(3) The technological development in telecommunications is so dynamic and full of new variations that legislation cannot keep pace. Until a law containing specific technical requirements has been passed on the European level, there are several new technical options which are superior with regard to one aspect or another and which can and should not be kept out of the market.

This regulatory dilemma seems insoluble, if a directive similar to the that on data protection in connection with ISDN as the main form of regulation is considered. Therefore, it seems necessary to look at different organisational and procedural options. A regulatory structure which in many aspects seems to be just the opposite of such a detailed specification of functions and features by law is the system of Public Utilities Commissions (PUCs) or Public Services Commissions (PSCs) in the United States. On the state level, by constitution or law, these commissions regulate public utilities which usually operate in a monopoly or at least in restricted competition. The laws mainly deal with the structures of these organisations and the procedures.

For example, in California tariffs for telecommunications services have to be authorised by the California Public Utilities Commission. Service providers have to submit an application for new services or features. An administrative-law judge decides whether an application is of public interest. If so, he or she can initiate a public notification and hearing. In addition, the judge can consult the Division of Ratepay Advocates where consumer lawyers, technicians, social scientists, etc, work. At the end of the procedure, a commissioner can either pass an application, reject it, or pass it with special conditions. In the case of calling-line identification there have been many hearings by PUCs and PSCs in the States.

This regulatory system is extremely flexible. It responds to new topics and concerns, such as privacy, which was not an issue ten years ago. It integrates the regulation of competition and prices with the regulation of service quality, including privacy and consumer protection. But this flexible reaction is – from a provider's point of view and an industrial-policy view – becoming a great weakness with the increasing dynamics of telecommunications technology. Eli Noam, professor at Columbia State University and for some years

Commissioner of the New York Public Services Commission, initiated an inquiry in order to set up principles of privacy. His main motivation was that the present procedures are not predictable for producers and providers. The Commission responds to services and features that usually are developed and ready for introduction to the market. If the Commission's response is negative or if modifications are required, new investments and development become necessary. As there are no general principles of privacy, the final decision depends entirely on the process of notification and hearings and is not predictable. Caller identification is a good example here, with a wide range of different decisions by the different commissions (see Figure 1). Noam lists more than 40 services and features of new telecommunications services which might become subject to similar processes. He argues that the US style of case law with regard to technical requirements is becoming ineffective and that the European approach of general rules and requirements leads to a more predictable situation for service providers and manufacturers. Therefore, he proposed a set of privacy principles (Noam) [14] [11] which were quite controversial when first discussed within the Commission, but were finally adopted in the overall policy statement of the PSC [15].

This example shows that the regulatory dilemma occurs also in countries like the US with a quite different regulatory system.

Figure 1.

2.2. *Dimensions of regulatory systems*

If the European countries are examined, it is found that there are still other organisational structures and procedural rules. Therefore, it seems necessary to undertake a systematic investigation of regulatory options, based on a survey of present forms in the area of data and consumer protection, and to try to learn from other subject areas, for example, information technology security (ITSEC). Without the resources of a large international research project it is not possible to carry out the detailed and comprehensive survey that is required. But a conceptual framework, containing the dimensions for such a comparison is proposed and illustrated by examples borrowed from different countries or subject areas.

Eight dimensions for describing and differentiating regulatory systems for data and consumer protection are proposed.

2.2.1. *General regulatory mode*

The main difference mentioned above between the ISDN directive and the US system may be called the regulatory mode. Two ideal types can be distinguished:

a) regulations containing requirements for telecommunications services (result-oriented);

b) regulations laying down procedures and authorising bodies which define these requirements (process-oriented).

The advantages and disadvantages of both types have been discussed in section 2.1. Of course it is possible to combine elements of both types.

2.2.2. *Specificity of requirements*

Requirements according to section 2.2.1.a) may be defined as:

a) (broad) principles stating what is to be achieved, but not saying how this has to be done;

b) detailed specifications, for example, in technical or organisational terms, or both, which lead to almost identical services.

The difference between the two options can be illustrated by the change of the rules on itemised billing in the EU directive. Originally, there was a detailed specification requiring that the last four digits of the called number be cut off, while the revised draft contains the principle that the itemised call statement shall preserve the privacy of calling and called users.

The same dimension can be applied to organisational and procedural rules, for example, the Public Utilities Codes in the US, the institutionalisation of the ombudsman in the Danish Payment Cards Act, or the expert groups according to the general EU data-protection directive. More specific rules can be found in the German data-protection law.

There are advantages and disadvantages. The more general rules allow a flexible adaptation to new technical features and new concerns, but they are less predictable for service providers and manufacturers.

2.2.3. *Authority of regulatory bodies*

If the mode of installing regulatory bodies is chosen, these:

a) may be consulted, may conduct inspections and issue reports, but without the power to make decisions;

b) may serve as a kind of ombudsman, to receive complaints and try to settle problems as they arise;

c) may give certifications with which service suppliers can demonstrate the degree of data or consumer protection, or both;

d) may decide on admission to the market and have the right to impose conditions.

a) An example for the first option is the German Federal Data Protection Commissioner who has to be consulted by Deutsche Telekom, and who conducts inspections and comments on the violation of privacy rights of people who are called in his biannual report. But he is not authorised to intervene in the process of data processing. Compared with the German situation, France Télécom has to consult the Commission Nationale Informatique et Libertés (CNIL) and reach agreement, which according to the scale employed, is authorised to grant admission to the market.

b) Ombudsmen act on request after problems have been raised. They are found for consumer protection in the Danish Payment Card Act or as the arbitration commission on complaints about telephone services in the Netherlands. The latter was set up by PTT Telecom, PTT Post and the two major consumer unions. It has a neutral chairman. One member is appointed by the PTT and one by the consumer organisations. The secretary who drafts the decisions has to be a lawyer (Korthals Altes) [16].

c) Certification indicates a process of acquiring a document which shows that a service or product conforms to a set of criteria that has been formally evaluated. Such certificates may only be provided by authorised bodies. They may either be a precondition for admission to the market or they may be acquired on a voluntary basis, assuming that users will prefer the services with certificates as opposed to those without. Examples of this option are to be found in the more technical aspects of information and communications technology. In particular, with regard to technical compatibility and security, the CEC has installed such systems (see for example, Höller) [17]. Conformity to compatibility standards is usually a precondition for the admission of terminals. Conformity testing has to follow specified rules, but may be conducted by private or semi-public organisations with special authorisation. The certificates then have to be submitted to the admission bodies or have to be shown only on request. A voluntary certification has been introduced for security [18]. Security in terms of integrity, confidentiality and availability is a basic requirement from the point of view of owners and users of technical systems. But to guarantee this requirements may require technical and organisational measures which might be quite costly. However, in different contexts security requirements might vary. Therefore, criteria are specified in order to distinguish between different degrees of security. Producers can decide which level is to be met, have the product evaluated by an authorised organisation and then get a certificate. This can be used for marketing the product. The customer can choose the desired degree of security and rely on the certificate.

This approach might also be applied to services with regard to data and consumer protection. For example, basic requirements might be defined as a precondition for offering a service and then different degrees of higher-level protection might be certified on a voluntary basis. Initially, state authorities were responsible for the criteria, the evaluation procedures and the authorisation of certification agencies. These agencies might be semi-public or even private organisations.

d) Finally, regulatory modes may decide on the admission of new services. The PUCs or PSCs in the United States and the French CNIL have already been mentioned as examples. Of course this is the strongest form of intervention into a market, raising great conflicts between economic objectives and rights. The legitimacy of this option rises, if the other forms have been tried and proved ineffective.

2.2.4. Legitimacy

Principles and procedures may be formulated, decided and put into operation as:

a) a law by parliament
b) a directive or ordinance by state authorities

c) a code of conduct or recommendation by some kind of association (self-regulation)

Most of the examples mentioned so far fall under categories a) or b). In Germany, the constitutional court recently rejected the privacy ordinance of the Minister for Post and Telecommunications. Because it restricts the constitutional right of privacy of communication, the parliament itself has to rule this restriction in detail by law.

Option c) is often offered by the industry which is to be regulated. Examples are payment cards or audiotext and videotext. In Germany, there is a self-regulating body for audiotext, and the European Information Industry Association (EIIA) recently presented 'Guidelines for Cross Border Control of Audiotex and Videotex Services in Europe'. According to the EIIA these guidelines should become part of the contracts between network operators and service providers. Thereby the guidelines would be more binding than in many other cases. However, the Legal Advisory Board to the CEC has stressed two problems. One relates to legitimacy, because the guidelines 'are drawn up by special groups which do not necessarily include all those involved, for instance users' (LAB, p 7) [19]. The second problem concerns compliance with national and Union laws. If the guidelines are made binding, they restrict market entry and operation, and this could be authorised only by law (ibid, p 8).

2.2.5. *Legal and institutional context*

The question of whether data and consumer protection in telecommunications services should be regulated within the legal and institutional context of market-regulation for telecommunications or in the area of the transactions concerned has already been raised. There are at least two alternatives:

a) any telecommunications service is regulated in the telecommunications context;

b) basic or reserved services are regulated in the telecommunications context, while value-added network services (VANS) are regulated in their respective fields.

The German ordinance on privacy in telecommunications (TDSV) follows option a) in some aspects. It contains requirements for the telephone service as well as some for videotext and e-mail. But at the same time it is addressed only to network operators and does not cover the relationship between the service providers and their customers.

The CEC seems to follow option b). The ISDN directive has been reduced to basic services by the elimination of Article 16 concerning teleshopping. Teleshopping will be regulated together with other forms of distant-selling in a special directive.

This option seems much more workable. But there are at least two problems. The first one concerns the boundary between the two categories of services. In the general regulation debate it has become obvious that there is no absolutely clear and lasting boundary between basic services and VANS. The second problem concerns coordination between the area of telecommunications and the area of application. The regulation of special services has to be enforced in the telecommunications area. Therefore some procedural link is necessary. It is more difficult to identify the need for regulation and to integrate different kinds of knowledge, which is necessary to come to sound rules. This is only possible by cooperation between telecommunications experts and experts, for example, in financial services, together with experts in consumer and data protection. But such cooperation in bureaucracy meets a lot of barriers. And if the rules are to cover any form of payment services, unsolved problems with, for example, off-line cards, might delay regulation of on-line electronic funds transfer at point of sale (EFTPOS). Therefore, there is a high probability that many processes for developing regulations are started but that almost none of them produces a result.

2.2.6. *Integration of subjects*

Questions of data and consumer protection are often complementary, but sometimes they conflict. In most countries (as has been shown for the EU), there are different areas of law, with different institutions. Coordination and cooperation are unusual. But from the point of view of the providers and users of telecommunications services it would be best, if all the relevant regulations were collected in one document, independent of the area of law they come from. And, from the point of view of the development of rules, there are advantages if the different issues are somehow related to each other.

There are at least four options:

a) co-existence of separate bodies and procedures, ad hoc resolution of conflicting demands, living with contradicting rules, or blocking of any rule at all;
b) creation of a new coordinating body which consults the different bodies responsible for the questions concerned;
c) integration of representatives for the different legal areas in one new institution which can make decisions itself;
d) creation of a new institution with new authority.

Option a) is what we have today. It is the default option, if b), c) or d) cannot be reached. The Public Services or Public Utilities Commissions in the United States are an example of option d). They integrate economic regulation with other issues of consumer protection and with the protection of privacy. This does not cause problems, because there are no other privacy authorities on the state level in the US.

As we have these institutions in most European countries, we can either organise some form of cooperation and coordination (options b) and c)). Or a new institution could be created, with authority for all issues related to telecommunications which takes away the authority of existing organisations. In some cases this might be an advantage, because responsibilities are dispersed and existing institutions are busy in so many areas that they deliberately pass on the responsibility for telecommunications.

If requirements are defined, as principles or very specifically, data and consumer protection should be integrated in the same law or directive. In Germany, this has been achieved by the videotext state treaty which contains regulations concerning privacy, advertising, price information, and many other issues.

2.2.7. *Dynamic adaptation*

As technology changes, new generations of software follow rapidly, offering new features and, in the market for VANS, providers still have to find out what customers want. The field that has to be regulated is highly dynamic. Therefore, result-oriented rules have to be adapted from time to time, and process-oriented approaches need periodic examination as to whether the bodies and procedures are still capable of dealing with the problems that arise. There are at least four options for such a dynamic adaptation:

a) laws valid for a limited time
b) consultative boards to the legislator
c) general principles with clauses for modification according to a defined procedure
d) relying on the existing adaptability of the legislator

Option a) is the radical answer to the experience that legislators are quite reluctant to adapt existing laws that lose their protective power with regard to data or consumer protection. But it has at least three disadvantages: it means uncertainty for network operators and terminal manufacturers who cannot anticipate what the next law will require; it means more work for the legislating bodies; and the legislator must decide in advance when the present law will expire.

This last disadvantage is overcome by option b). A consultative board monitors the development and makes recommendations to the legislator. This option is employed by the CEC in different fields. The general directive for data protection, as well as the special one for ISDN, contain the requirements for an advisory committee, composed of the data-protection authorities of the member states, which is to be consulted by the Commission before any changes to the directive. They are also required to take the initiative and make recommendations for improving the protection of privacy (Arts 31 and 32 of the proposed directive) [6]. Another example is the Legal Advisory Board which has been established within the IMPACT programme and which is consulted on questions of legal barriers to a single information market.

Option c) tries to solve the conflict between the appropriateness and predictability of regulatory requirements by leaving modifications to the defined procedure where, for example, the parties concerned may have a say. Thereby the general principles allow for basic predictability, while smaller adaptations can be made without legislation.

2.2.8. National and international level

As the economic and political goal is to build a single European market for telecommunications services, offering economies of scale to providers and a greater choice to users, harmonisation is necessary within the EU.

Laws and bodies may be enacted:

a) on the EU level only
b) on both national and international levels
c) only on the national level according to harmonised standards
d) only on the national level without harmonisation

Option a) could only be applied to certain issues. There would need to be additional national regulation until the 'United States of Europe' is created – which is becoming less likely.

Option b) would be a combination of the same framework regulation and minimal standards at the EU level, with the option for more detailed procedures and additional requirements within this framework. But it might also be useful to have interworking bodies on both levels, as for instance in technical standardisation.

Option c) could be realised by directives or recommendations addressed to the member states to adapt their national legislation to common standards. Experience so far shows, however, that there are always some members who do not respond – and then the objectives of harmonisation are not met.

Option d) is what we have. If there is no harmonisation, the differences will increase even further in the future because of new providers entering the markets, new technological features, etc.

2.3. Elements of one possible regulatory system

In order to reach a more detailed discussion, the following example is suggested as one possible selection out of the regulatory 'tool-kit' presented above.

The basic idea is that there should be a distinction between the basic telecommunications services that are used by millions of people and more specialised VANS. The basic services should be regulated in detail on the EU level, for industrial-policy reasons as well as from a human-rights perspective.

These mass-services affect the human rights of almost every citizen in the Union, and provide the biggest markets where large economies of scale could be realised, if there were identical technical requirements for terminals, switching technology, accounting software, etc.

2.3.1. Basic services

Because the basic human rights of nearly every individual in the Union are concerned, a directive is the only appropriate form. In order to provide a maximum of certainty for users, providers and manufacturers, the requirements should be specified in more detail than in the present draft directive. There should be no doubt about which services and providers fall under this directive. The telephone service, fax, e-mail, and video-phone services should explicitly be incorporated. And any network operator or service provider offering these services should also be included.

With regard to the deregulation of the telephone monopoly in 1998, there should be an obligation to create a national regulatory organisation which will be responsible for licences for these services, for determining that these are conforming to the requirements, and which may also be consulted in advance by service providers. In addition, there should be a committee for complaints.

The necessity for licences for basic services, the procedure for admission, the responsibilities and structure of the organisation should be laid down in national telecommunications law. The regulatory organisation should integrate economic and competitive supervision together with data and consumer protection. Existing organisations should be coopted onto the board or there should be special cooperation agreements, or both, even sending officers from existing organisations to the new one. This fundamental change, which makes strict regulation necessary, will at the same time make it easier to build up new organisations.

The dynamic adaptation to new technical-service features or even to the incorporation of new services (for example, pay-per-view television via the telephone network) might be achieved by the advisory committee which could be built up according to the existing draft directive.

2.3.2. Specialised VANS

Specialised VANS will only affect subgroups of people. They will raise different problems depending on their content and context, and there will be a dynamic development of new services, with new service providers emerging, although some will disappear. Therefore, a result-oriented mode of regulation could not cope with the complexity and dynamics in this field. Legislators cannot leave the development completely to market forces. They are obliged to guarantee the privacy of communications and individual rights. And in order to create markets as large as possible, harmonisation is necessary.

The overall regulatory framework should be composed of a process-oriented framework directive in the telecommunications context and of special result-oriented and process-oriented ones for certain areas such as distant selling or electronic payment. The framework directive should contain basic principles and minimum requirements, as well as criteria and procedures for determining different levels of additional data and consumer protection.

On the national level, there should be authorised agencies for certification, and regulatory organisations responsible for ensuring that minimum standards are met, and that services are in line with certificates. They should have inspection rights and be able to deal with complaints. The main emphasis should be on market forces, because consumers can choose between different services with different levels of protection.

The regulatory body, which could be the same as that for basic services, should integrate the different aspects, which in this case include general data protection and consumer law as well as special EU and national legislation on the content of some services. It will need to be a statutory body, with its establishment based partly on telecommunications legislation.

On the EU level there is a need for a coordinating body to monitor developments, make suggestions and support cooperation between those from the telecommunications area and those from the application areas, among users, the Commission, the European Parliament and national bodies.

This whole framework should be laid down in EU directives, in particular in the existing one on competition between telecommunications services and the new framework directive suggested here. National legislation should specify the general principles.

3. Final remarks

These suggestions need to be elaborated in more detail. There are also arguments for other options on each of the dimensions. However, researchers cannot dictate the system. At the end of a conference held in Berlin in February 1991, where representatives of data-protection authorities, consumer organisations and universities met for the first time, participants agreed on a memorandum asking the CEC and the European Parliament to initiate an action programme on integrated data and consumer protection in order to continue the discussion on a permanent basis. Unfortunately this request was not met.

Without such a programme combining state-of-the-art surveys, elaboration of regulatory frameworks and scenarios and workshops involving the different groups concerned, there will be no progress.

It is not clear why progress is so slow. Is it that the most powerful political actors do not see the need? Or is it mainly because there are no means of bringing together people from such different areas as the information industry, consumer protection, economic telecommunications regulation, and those concerned with privacy?

It is disappointing that the draft ISDN privacy directive has not yet been passed by the Commission. On the other hand the European Information Industry Association has submitted guidelines and is seeking the cooperation of the Commission. The main motivation is the fear that certain audiotext services might produce a poor image of the whole market. Some kind of regulation will improve customer-confidence. (LAB, p. 7) [19]. Initiatives such as this might create a more favourable climate for regulation. In any case, a new regulatory framework has to be implemented by 1998. Without an equivalent to the protection of the privacy of communications by public network operators, constitutional courts will not admit private telephone companies.

This paper still leaves open a lot of questions. It is incomplete with regard to the problems and the approaches taken so far to regulate different services in the member states. The systematisation of institutional options is both incomplete and too general. It is not just a matter of weeks or months that are required to overcome these deficiencies. Considering how long it takes to get from a comprehensive study to a directive implemented in national law, 1998 is not far away. Meeting this deadline is quite a challenge. Furthermore, the next big issue is already on the horizon. If multimedia technology, and in particular data-compression technology, is advancing as it has in the past, pay-per-view television will be available in the telephone networks. This will raise the question of whether this is television in legal terms and whether the existing regulation of this sector applies.

While European politicians often refer to the US when pursuing their deregulation policy, it is interesting to note that the situation there is now changing. In the Action Programme of the Clinton-Gore initiative for a National Information Infrastructure (NII) the issues of privacy as well as the integration of the different regulatory systems for telecommunications and television have been given high priority. And it is stressed that these issues have to be dealt with in such a way that all the different interest groups in society can participate.

It is hoped that Europe keeps on learning from the US. Of course telecommunications in Europe could be completely deregulated, and then reregulated on the basis of the experience. Or the ongoing debate within the NII about universal service, public interest applications, privacy, etc, might be examined with a view to incorporating some elements already in the continuing process of re-structuring European telecommunications.

Annex

A.1. Problems of data and consumer protection related to the telephone service and network operators

Problems of data and consumer protection related to the telephone service and network have received most attention, because nearly everybody is concerned. Up to now, three main problem areas have been discussed, two of which are explicitly regulated in the draft of an ISDN data-protection directive.

A.1.1. Directories

Up to now, the privacy aspect of directories has mainly been discussed in terms of the right to have one's telephone number listed or not. In November 1989, an expert group on data protection of the Council of Europe's 'New technologies: telecommunications sector' presented a summary of the problems and the approaches taken by the member states [20].

Some member states make the listing of numbers in directories a precondition for participation in the telephone service. For example, in Germany a subscriber at that time had to justify the wish to be excluded from the telephone directory by demonstrating that inclusion might give rise to distress or discomfort. In other countries (for example, Austria, France, Spain, and Norway) subscribers have a right not to have their phone numbers listed; France Télécom, however, charges a fee for this.

The experts state that there are great differences between the types of personal data required by PTTs. For example, in Germany subscribers have to give their date of birth, and in Norway their personal-identification number. While in some countries subscribers may enter additional data, Austrian law rigidly defines which personal data may be collected.

Another issue of concern is the commercialisation of subscriber data. In some countries, for example, France, there are lists to which subscribers may add their names, if they do not want their data made available to third parties.

Although the experts have made no explicit recommendations, reading between the lines, they appear to favour:

- the right not to have one's phone number listed, free of charge;
- restriction to only those personal data that are necessary for providing the service, but with the individual option for additional data;
- the right of subscribers to be informed about the commercial policies of PTTs and the right to opt out of the release of one's own data to third parties.

The draft ISDN directive of July 1990 did not include such regulation. However, the European Parliament added a paragraph which has been incorporated into the revised draft of DG XIII, dated October 1992 (see Article 4 below). In order to control these rights, subscribers are entitled in Article 6 to obtain confirmation as to whether personal data are stored.

However, during the past few years, the technology and organisation of telephone services, networks and other services have changed dramatically:

(1) The network operator and the service provider are no longer necessarily the same. For

example, in mobile-communications networks in Germany, potential subscribers have a choice between nearly a dozen service providers who themselves buy access cards from two, and in the near future three, network operators. For billing and security purposes, some data on subscribers have to be exchanged between service providers and network operators. But which data have to be exchanged between whom and for what purposes?

(2) The network operator or service provider, or both, no longer has the exclusive right to edit directories. Rather, directories – be they in printed or electronic form – are becoming a service in themselves. International standardisation bodies have developed technical standards for electronic directories in order to overcome technical barriers to their distribution, combination and remote access to them. Should subscribers have the right to decide in which directory they want to be listed? Should they decide, for example, to be in the directory of their telephone-service provider, but not on a CD-ROM which might also provide for reverse access? And as there may be providers of directories that subscribers do not know about, how could they exercise their right to control these data? Should a provider of directory services be obliged to obtain the consent of the subscribers concerned?

(3) With regard to consumer protection all this will raise problems of quality control. In Germany, for example, Telekom is considering charging for information services. But as the probability increases that directories are not up-to-date, the question arises whether you have to pay if you get a wrong number or if you are told that the person is not in the directory.

All this is to say that even the revised draft needs to be adapted to the new, more complex situation that is evolving in the field of directories.

Table 1. Article 4

Original Draft (July 1990)	Revised Draft (October 1992)
	Principle
1. Collection, storage and processing of personal data by a telecommunications organisation is justified for telecommunications purposes only, in particular in order to establish connections for the transmission of voice, data or image, to produce bills, to compile directories, and for other legitimate operational purposes, for example, fault clearance, prevention of misuse of the telecommunications organisation's equipment, or registration of incoming calls in accordance with Article 13 (1).	1. The processing of personal data by a telecommunications organisation is justified for telecommunications purposes only, in particular in order to establish connections for the transmission of voice, data or image, to produce bills, to compile directories, and for other legitimate operational purposes, for example, fault clearance, prevention of misuse of the telecommunications organisation's equipment, or registration of incoming calls in accordance with Article 13 (1).
2. The telecommunications organisation shall not use such data to set up electronic profiles of the subscribers or classifications of individual subscribers by category.	2. Personal data contained in a directory should be limited to what is strictly necessary to identify a particular subscriber, unless the subscriber requests additional personal data to be published. The subscriber shall be entitled free of charge, not to have his or her sex indicated and to be omitted from the directory at his or her request.
	3. The telecommunications organisation shall not use such data to set up electronic profiles of the subscribers or classifications of individual subscribers by category, for purposes other than those mentioned in paragraph (1), unless the subscriber has given his consent.

A.1.2. Itemised billing

Itemised billing raises questions of privacy on two levels, which leads to conflicts in the requirements for consumer protection.

The first level concerns the storage of traffic or connection data [21] for each single call in order to produce an itemised bill regularly or only on request, if the summarised bill is questioned. Here, the main concern for privacy is that state authorities or commercial organisations might get access to these data without the consent of the subscribers concerned. In legal terms, it is a matter of the privacy of communication which, in many countries, is a constitutional right which may only be broken in special cases, in particular in the pursuit of major crimes or for the sake of national security. In order to resolve the conflict between the right to unobserved communication, as it has been called in Germany, and the duties of police authorities, secret services, etc, to pursue crimes, specific procedures were developed, such as the requirement for an order by a judge. But, according to an international comparison by Garbe, only Germany, the United Kingdom and the United States have regulated the interception of telephone calls by law (Garbe, p 84) [22]. In Germany, civil rights groups have complained that the respective law was written in 1928 when there were no data held on telephone calls, so that it only allowed for interception of those calls after an order by a judge and that, meanwhile, the situation has changed. Today, data are available on calls spanning several months. In addition, the complaint has been made that the obligation of Deutsche Telekom to reveal these data is not confined to cases of major crime.

There is a consensus that the privacy of communications is covered by the European Declaration of Human Rights. Yet, neither the EU draft of the special privacy directive nor any other directive regulates the conditions under which network operators must reveal connection data to state authorities, despite the fact that with increasing international connections, subscribers of other member states may be involved.

A general regulation seems overdue, if networks and services are to be run by private organisations. At least in Germany, basic rights apply directly only to the relationship between the individual and the state authority, but not between individuals. Yet users of telephone services must be able to have trust in the privacy of communications regardless of the legal status of the network operator.

The second level of privacy concerns the bill itself which discloses the communication between A and B, and between A and other people in his social environment. A's interest in checking his telephone bill may come into conflict with his wish not to reveal certain calls to other people in his surroundings. If he wants the full number called on the invoice, this remains incompatible with the right to privacy of communication or informational self-determination of the person called, or both.

While itemised bills are given out in Belgium and, for example, in the US, without concern for privacy, the issue was raised in France back in 1985 (Gebhardt [23]; Garbe [22]). According to the French privacy law of 1978, France Télécom had to consult the Commission Nationale Informatique et Libertés (CNIL). After several years of negotiation, CNIL ruled that the last four digits of the called number be deleted on the bill, that the full numbers be kept by France Télécom, and that subscribers be allowed to have a look at these data, but not to get a copy.

This approach to resolving the conflict between data and consumer protection, and the rights between the calling and the called party, has served as a model for regulation, for example, in Switzerland and Germany, and for the draft of the EU directive. However, many commentators do not feel that this is a sound compromise. British Telecommunications officials and the data protection registrar in particular are rejecting the EU draft, especially because of this point. Meanwhile, other technical procedures that have been suggested as possible solutions, have been either planned or undertaken.

In the Netherlands, PTT offers the choice between a summarised and a categorised bill with subtotals for local, national and international calls, audiotext services, videotext, etc. (see Korthals Altes) [16]. This way, subscribers can ask for itemisation of each category. As the charge for each item is five Dutch cents, consumer organisations recommend that local calls be summarised and only national, international and other services itemised.

In order to acknowledge the right of the called person, unlisted numbers are generally not shown and any other subscriber may request that his number is not entered on the bills of people who called him. This takes into consideration help-lines and other organisations which need to preserve their clients' anonymity.

The Dutch solution resolves a problem which deleting the last three or four digits does not resolve, and which, in Germany, has led to a delay in introducing itemised billing, despite the regulation of data protection in 1991 [24]. The German ordinances stipulate that calls to help-lines and other institutions providing consultation on social and religious matters, and which are bound to confidentiality, may not be revealed by the bill. As the first digits of some help-lines are the same all over Germany, the deletion approach does not help in every case. If the number called is not shown at all, this might start talk about who has something to hide, for example, in the family. And if the whole data set for such a call is left out, there will be a difference between the amounts for individual calls and the sum of the bill. Therefore, a subtotal is needed in which the calls to help-lines, etc, may be hidden. These may be local calls, for which no strong argument of consumer protection can be put forward, in particular if the listing of each call has to be paid for separately. However, this does not work in mobile telecommunications where, at least in Germany, there are no local calls. Here, the Dutch solution resolves the problem because a 'masked number' may be that of any person who has not given consent to have his number listed. However, officials of Deutsche Telekom and of Mannesmann Mobilfunk argue that their computer software for billing is not able to mask certain numbers and that they cannot find software producers who would add such features.

The draft of a special EU directive originally adopted the French approach. This has been accepted by the European Parliament. But then the Dutch did not want their solution to be excluded. Therefore in the revised draft of DG XIII a more general formulation has been chosen.

On the one hand it seems wise not to fix details in a directive which cannot be changed easily A lot of other options were suggested in the debate on itemised billing. The expert group of the Council of Europe suggested the application of prepaid chipcards, and in Sweden, according to Garbe, Televerket plans to offer a per-call control of billing by pressing a button before the respective call (Garbe, p 50) [22].

On the other hand, if legislation is not very specific, it leads to the implementation of different technical options, for example, for billing-software packages, and does not promote the technical harmonisation which, with regard to economies of scale, is one of the industrial objectives of the directive.

The issue of calls to help-lines shown on an itemised bill is not covered by the draft at all.

Table 2. Article 11

Original Draft (July 1990)	Revised Draft (October 92)
Upon application of the subscriber an itemised call statement may be produced containing among other items the telephone numbers of the called subscribers without the last four digits.	Member states shall ensure that upon application of the subscriber an itemised call statement is produced. It shall preserve the privacy of calling users and called subscribers.

A.1.3. Calling-line identification and other features

Another hot issue in many countries is calling-line identification. Because of digital switching technology and a new signalling system it has become possible to send the number of the calling line to a display on the telephone at the called line. In some countries, this new signalling system has been introduced together with the ISDN technology, and the debate has become one about problems of the privacy of ISDN. In other countries, for example, France and the US, the signalling system has been introduced together with the digitalisation of the switching technology.

Developers of the new system thought it a matter of politeness that calling parties identify themselves if they intrude into the homes of others. Often the analogy is made to the peephole in doors through which the owner of a flat can check who is ringing (see, for example, Katz) [25]. However, the basic difference is that calling-line identification depends on the equipment of the called party which the calling individual cannot see. This means that persons calling do not know whether their number is shown to the called party. In this context, a conflict of privacy is seen between the right of the called person and the right of the calling person to control the data relating to him.

In Germany, the debate was not as heated as, for example, in the US, because Deutsche Telekom confined the calling-line identification to calls between ISDN lines at a very early stage. In the US, however, anyone could purchase a small box which showed the numbers of calling lines. Thus, everybody had to take into account that their number may be shown to the called party. As telephone companies charged for this box, they had to obtain approval for this feature by the regulatory bodies at the state level. And there, a long series of protests and interventions came up. In hearings, court trials and legislative action a variety of arguments were presented which went beyond the simple privacy conflict between the rights of individuals. In many cases, the called party is an organisation which cannot claim privacy concerns for itself (Rotenberg, p 67) [26]. In most of these cases, identification does not take place in direct face-to-face communication, as it would if you go into a shop and ask for the price of a product. And as commercial organisations have the means not only to see, but to store the numbers of the calling lines, the right to control whether or not one's own phone number is passed on has, meanwhile, been considered of greater importance by nearly all regulatory bodies (for an overview see Slaa [27]; Shultz [28]; *Privacy Journal*, November 1992 and July 1993).

The first technical solution to this problem was demanded by the French CNIL which ruled that France Télécom may introduce calling-line identification only if they offered a blocking facility for the calling parties at the same time. And as the context and requirements regarding identification and anonymity vary from call to call, CNIL asked for a per-call blocking facility. Only because this was not immediately feasible, a per-line blocking was accepted for some time (see Gebhardt [23]; Garbe [22]).

Per-line blocking was admitted very quickly in every country, because subscribers with unlisted numbers, among whom are the most influential politicians, did not want to reveal their private numbers. Per-call blocking was controversial in many countries but finally became the rule, thus shifting control to the calling party. In order to provide for the protection of called parties against intrusion in general or against obscene calls, etc, it has been suggested that a blocking facility against unidentified calls be introduced for called parties.

The draft of the EU directive includes per-call blocking as well as the blocking of unidentified calls.

Table 3. Article 12

Original Draft (July 1990)	Revised Draft (October 1992)
1. With regard to communications between subscribers linked to digital exchanges the calling subscriber must be able to eliminate, via simple technical facility, the identification of his or her telephone number on the display of the called subscriber's terminal equipment, or its recording in a storage facility of this terminal, on a case-by-case basis. The transmission of the telephone number may also be permanently eliminated by the telecommunications organisation upon application of the calling subscriber.	1. With regard to communications between terminals linked to digital switches, the calling subscriber must be able to eliminate, via a simple means and free of charge, the identification of his or her subscriber number on the called subscriber's terminal equipment, or its recording on a storage facility of this terminal, on a per-call basis. It must also be possible for the transmission of the subscriber number to be permanently eliminated by the telecommunications organisation upon application of the subscriber.
2. The called subscriber may apply for permanent elimination of the identification of all incoming calls; he or she must also be able to turn off the display of his or her terminal equipment, or to eliminate the recordings in the terminal's storage facility, in order to prevent the identification of the incoming calls, on a case-by-case basis. The called subscriber must be able to limit the acceptance of incoming calls to those which identify the calling subscriber's number.	2. The called subscriber may apply for permanent elimination of the identification of all incoming calls. The called subscriber must be able to limit the acceptance of incoming calls to those where the identification of the calling subscriber's number has not been eliminated. 3. With regard to terminals linked to an analogue switch and terminals linked to digital switches and as far as full application of paragraph (1) is not possible, the former subscriber is to be informed of the identification of his or her subscriber number and to be offered free of charge, the choice between the per-line and the per-call elimination of the feature upon application.
3. With regard to communications between a subscriber linked to an analogue exchange and subscribers linked to digital exchanges, the former subscriber is to be informed of the identification of his or her telephone number and to be offered the permanent elimination of the feature upon application. The subscriber must also be able to eliminate the identification on a case-by-case basis.	4. The provisions set out in paragraph (1) and (3) shall also apply with regard to calls to third countries.

Telecommunications manufacturers, and network operators in particular, opposed the regulation concerning subscribers connected to analogue switches, because their information and the provision for eliminating the forwarding of their phone numbers meant additional technical effort which so far has not been planned for in the European switching systems. But, as it is the rule in the US and apparently because there is a lot of pressure, the Commission largely stuck to its original proposal. The alternative would have been to confine calling-line identification to calls among ISDN subscribers. But because ISDN is growing very slowly in residential areas, this would mean that for all calls coming from non-ISDN subscribers, the terminal could not identify the calling line. And then subscribers would not be willing to pay an extra rate for the identification of only five or ten per cent of all incoming calls.

With regard to other features, which were not subject to such heated discussion, the commission did not stick to its original technical suggestions.

The original suggestions concerning call forwarding required technical means to take account of the rights of third parties to whom calls are forwarded, and of the calling subscribers whose calls are forwarded. Already in the compromise reached between the different committees of the European Parliament, these specific requirements have been transformed into a vague requirement for the development of technical means.

Table 4. Article 13

Original Draft (July 1990)	Revised Draft (October 1992)
1. For a limited period of time, the telecommunications organisation may override the elimination of the calling-line identification: (a) upon application of a subscriber requesting the tracing of malicious calls. In this case, the data containing the identification of the calling subscriber will be stored by the telecommunications organisation and be made available upon request to the public authority charged with the prevention or pursuit of criminal offences of the Member States concerned; (b) upon specific court order, in order to prevent or pursue serious criminal offences. 2. A permanent override function must be made available upon request: (a) to organisations recognised by a Member State which answer and deal with emergency calls; and (b) to fire brigades operated or recognised by a Member State. 3. The telecommunications organisations shall take the necessary steps to ensure that the override function is operational on a national and Community-wide basis.	Exceptions 1. Member States shall ensure that, for a limited period of time, the telecommunications organisation may override the elimination of the calling-line identification in exceptional cases: (a) upon application of a subscriber requesting the tracing of malicious calls. In this case, the data containing the identification of the calling subscriber will be stored by the telecommunications organisation and be made available upon request to the public authority charged with the prevention or pursuit of criminal offences of the Member States concerned; (b) upon specific court order, in order to prevent or pursue serious criminal offences. 2. Member States shall ensure that a permanent override function is made available upon request: (a) to organisations recognised by a Member State which answer and deal with emergency calls; and (b) to fire brigades operated or recognised by a Member State. 3. The telecommunications organisations shall take the necessary steps to ensure that the override function is operational on a national and Community-wide basis.

Table 5. Article 14

Original Draft (July 1990)	Revised Draft (October 1992)
1. Calls may be forwarded from the called subscriber to a third party only if this party has agreed; the third party may limit automatic forwarding to those calls which identify the calling subscriber's number; the third party must be informed via a specific signal of the message that the call has been forwarded. 2. The calling subscriber must be informed automatically during the establishment of the connection that the call is being forwarded to a third party.	Calls may be forwarded from the called subscriber to a third party only if this party has agreed; for this purpose ways and means of agreement by a third party will have to be developed and provided for as well as the possibility to stop automatic forwarding.

However, for digital private branch exchange systems (PBX), the technical procedures originally requested have now been implemented, at least in Germany, because shop stewards in particular exercised their right of co-determination. Other features which are subject to internal regulation are (see Höller and Kubicek [29]; Roβnagel [10]):

- Conference calls: participants who join a conference should be informed by their display that they are entering a conference.

- Incoming-call log and automatic callback: participants should know if incoming calls are registered and be able to get off this list, in particular if the listing would lead to an automatic callback after the return of the called party.
- Direct speaking into other rooms: because participants with this feature can turn on the loudspeaker and the microphone of other terminals in other rooms, this feature also presents a means of silent surveillance of rooms. The requirement is that the person speaking into the other room can turn on the loudspeaker, but that the microphone only be turned on by the person addressed.
- Hands-free and open-listening operation: with the provision of loudspeakers and microphones in more and more telephone terminals, people in offices use these facilities in order to keep their hands free for handling documents, taking notes, etc. Some deliberately make the content of calls available to other people in their room or they have their office door open and everyone passing can listen. The requirement here is that turning on the loudspeaker has to be signalled by a beeping sound or light to the other party.

It is astonishing that this last point has not been subject to similar public concern as that caused by calling-line identification. Because we are concerned here with breaching the privacy of communication, not only with regard to who is calling, but with the content of the communication as well. And there are many cases where we would not say something if we knew that someone else was listening. One explanation is that, for some time, it has been accepted in business that calls were recorded. But even more important is that the police and secret services who are eager to wiretap do not want too specific a regulation.

The original draft of the EU directive concentrated on the relationship between two individuals and asked for the informed consent of the other party. But, in the revised draft, the obligation of member states has been incorporated to regulate surveillance of communications. And at the same time the requirements regarding the obligations of subscribers were formulated more vaguely.

With regard to telephone conferencing, which has now been introduced with the implementation of the Euro-ISDN standard, no regulation has, thus far, come to my knowledge. And features such as the incoming-call log, automatic callback, etc, are still in the international standardisation phase. Unfortunately, the officials there do not seek discussions with data-protection authorities in advance.

Table 6. Article 15

Original Draft	Revised Draft
	Surveillance of communications
1. If the content of telephone calls is made accessible to third parties via technical devices such as loudspeakers or other on-hook equipment, or stored on tape for own use or use by third parties, provision must be made in order that the parties concerned are informed via an appropriate procedure of such diffusion or storage before the diffusion or storage is initiated and for as long as it continues.	1. Member States shall take appropriate measures to ensure that listening or tapping devices or other means of interception or surveillance of communications are applied only subject to authorisation by the competent judiciary or administrative national authorities. Member States shall ensure that the content of telephone calls must not be made accessible to third parties via technical devices, such as loudspeakers or other on-hook equipment, or be stored on tape for own use or use by third parties without the consent of the users concerned.
2. Paragraph 1 does not apply in the cases covered by Article 13(1).	2. Paragraph 1 does not apply in the cases covered by Article 13(1).

A.2. Problems of data and consumer protection in value-added network services

While the public debate and the regulatory response to problems of data and consumer protection with regard to the telephone service is rather well documented even by international comparison, information about similar problems with value-added network services is rather scattered. This may be due to the fact that the use of these new services is still rather low. But, on the other hand, the threats to privacy are greater because these services, by definition, imply the electronic processing of the content of communications and, in most cases, a registration of the services for billing purposes which allows for exposure of the content.

A.2.1. Overview

In 1989, the group of experts on data protection of the Council of Europe examined the rules laid down in the Data Protection Convention of 28 January 1981 to see whether they still provided protection against the threats emerging from the new technologies and new applications, particularly in the field of telecommunications. They examined three types of new services (Council of Europe, p 7) [8]:

(1) Telemetry: 'The remote collection of personal data by automatic means, in which the data subject plays no active part in the actual process of the collection of data.'

(2) Interactive media: 'Automatic data-processing systems which provide the basis for instantaneous dialogue between an individual and such a system with a view to obtaining and transmitting information at the initiative of the individual.' This category refers to videotext systems including their application to telebanking and teleshopping, online databases and even pay-per-view cable television.

(3) Electronic mail: 'A telematic system which enables person-to-person storage and forward communication such that, as required, messages can be sent and accessed by a central facility, the sending of the messages being under the control of the sender and their reception under the control of the addressee.'

An international comparison conducted by Gebhardt, which was one of the inputs to the draft of the special EU directive, showed that some regulation for videotext is quite common, while regulation for telemetry, e-mail and interactive cable television is still rather rare.

Table 7.

Service/features	CH	F	D	UK	S	USA	J
value-added services				0			
videotext	x[1]	x	x	x	x	0	
itemised call statements	x	x	x[2]				x
call recording and malicious-call tracing		x	x				
electronic mail		x	x	x			
TEMEX (telemetry)			x				
calling-line identification		x	x	x			
call forwarding			x				
interactive cable-television systems		x				0	

CH = Switzerland; F = France; D = Federal Republic of Germany; UK = United Kingdom; S = Sweden; USA = United States; J = Japan
x = law, regulation or instruction; 0 = guidelines or codes of conduct

[1] Draft data protection regulations have not yet come into force.
[2] Future regulations.

Table 8. Article 16

1. The telecommunications organisation must ensure that the telephone number as well as other personal data of the subscriber, in particular concerning the quantity and nature of his or her orders when using a teleshopping service, is stored only to the extent strictly necessary to supply the service and is only used by the service provider for purposes authorised by the subscriber.
2. Subject to the provisions of Article 20, the service provider may not set up electronic profiles of subscribers or classifications of individual subscribers by category, without their prior consent.

As a result of these findings the original draft of the EU directive contained an article on electronic profiles in connection with teleshopping services (Article 16).

However, due to pressure from the Committee for Economic Affairs, the European Parliament requested that this article be deleted completely. The official argument was that according to Article 19, the provisions of the directive should be applied to other public telecommunications services as well and, therefore, there was no need to deal with teleshopping in an extra article. At the same time, the European Parliament requested that the scope of the directive should be extended to service providers other than telecommunications organisations. This has been considered in the revised draft (Article 19).

Neither the original nor the revised text of Articles 18 and 19 is clear to me. According to the original text, the directive also applied to services like telex, teletext and e-mail which are supplied by telecommunications organisations and which present similar risks with regard to itemised billing and calling-line identification. The revised text only refers to services provided by other suppliers and does not cover the articles on itemised billing, calling-line identification, etc. Only if one understands the word 'call' in paragraphs 11 to 14 as covering telecommunication by means of fax, texts, data, or pictures, the new phrasing would be equivalent. An indication might be that the word 'telephone number' or 'telephone call' has been avoided in the revised draft. However, it is not explicitly stated to which services the regulation applies. Furthermore, the rules do not cover specific threats which stem from forms of communication other than speech.

Table 9. Article 19/18

Original Draft, Art. 19	Revised Draft, Art. 18
	Application to other service providers
1. The provisions of this Directive relating to the telephone service shall be applied to other public digital telecommunications services to the extent that these services present similar risks for the privacy of the user.	1. Articles 4 to 10 and 17 of this Directive shall apply mutatis mutandis to other public digital telecommunications services of service providers other than telecommunications organisations.
2. The measures necessary for the implementation of paragraph 1 shall be adopted by the Commission after consultation of the Working Party referred to in Article 22 and in accordance with the procedure laid down in Article 23.	2. The measures necessary for the application of other provisions of this Directive to service providers other than telecommunications organisations or measures that may prove necessary to give better effect to the application of paragraph (1) shall be adopted by the Commission and in accordance with the procedure laid down in Article 21.

However, it is quite obvious that the services mentioned pose problems of data protection that cannot be handled by the provisions of the draft directive on ISDN. In February 1991, a conference was held in Berlin that brought together representatives from data-protection authorities and consumer organisations from ten European countries. They did not only discuss the draft of the special directive but also explored the problems connected with services such as teleshopping, telebooking, electronic funds transfer at point of sale (EFTPOS), online databases and electronic mail (see Kubicek) [30].

One question that was raised by the discussion in Berlin related to the appropriate area of regulation, for example, of teleshopping and EFTPOS. Is this an original concern for telecommunications regulation or should it be part of the regulation of the respective area of transactions? In any case, in the following overview, recent regulations concerning distant-selling or electronic-payment systems will have to be taken into account.

A.2.2. E-mail

The basic structure of an e-mail system corresponds very closely to the structure of the traditional postal service with post offices.

Many companies have internal e-mail systems, and in almost every member state of the EU there is at least one public system. In member states with deregulated telecommunications there are several e-mail services. In the past, many of these were not connected. But gradually, they are becoming compatible due to the international standard, X.400.

According to the expert group of the Council of Europe, the main privacy concern relates to unauthorised access to messages stored in post-office computers. Usually, access is only possible with passwords. An additional means of security is encryption. However, it seems that state agencies oppose the introduction of encryption between user terminals because it prevents them from intercepting electronic mail.

The experts of the Council of Europe mention an even more fundamental question concerning the character of electronic mail: are they letters or files? The question becomes crucial if a letter contains personal data about a third person. If it is a letter, then the service provider must not look at the content and has to ensure confidentiality. If it is a file, data-protection law would have to be applied, including notification of the person concerned or at least allowing inspection by this person.

For security reasons, the experts ask for measures such as automatic dating and the dispatch of acknowledgement. They do not mention that with such features, privacy conflicts arise which are similar to those in the telephone service. If the rules concerning calling-line identification have to be applied, there should be the option to eliminate the data of the sender in the letterhead. But one could argue that there should be no anonymous letters in e-mail systems at all and that there are other ways of communicating in cases where anonymity is justified. But severe problems arise with the feature of message forwarding. If A sends a letter to B and B has activated a forwarding function to C, A would not have the chance – in contrast to a telephone call – to hold back the message, because e-mail is usually asynchronous. And C would have access to the whole message at once. This means that the risk of unintentional disclosure of the content of communications is greater than with telephone communications and transparency for the sender is weaker, ie, the originator of the message has no control over its disclosure. The problem arises also with telefax. However, telefax is viewed by most users as a semi-public communication channel, more comparable with post cards than letters. In contrast, e-mail is thought to be as confidential as letters.

To handle the problem of unintentional disclosure, in X.400 the sender can control whether his message is forwarded or whether it is sent back to him, if the destined receiver has activated forwarding (see Roβnagel [10] and Höller [17] for this and other features). However, there is no regulation that public e-mail systems have to apply to X.400.

Itemised billing, by the way, is not mentioned as a problem in any text on the privacy of e-mail. In addition, e-mail systems offer the possibility of asking for a receipt for each message which, in some systems, not only states the time when the message was delivered to the post box of the receiver, but also when it was activated by him. This does not necessarily imply that it has been read as well. In these cases, information about the receiver is produced by the system and passed to the sender without the knowledge and without the consent of the person concerned.

Although there are some similarities between the telephone service and e-mail, the examples show that rules concerning identification, forwarding, or registration should be differentiated.

The draft of the EU directive does not contain any specific regulation for e-mail services. In Germany, the privacy legislation for telecommunications contains one article concerning 'message handling systems with storage facilities', which covers e-mail as well as voice-mail systems and any other form of storage of messages. According to this rule, the network operators and service providers may store and process messages only according to the order of the client during his input, and only the client decides who has access to his messages.

The network operator or service provider may notify clients that receivers have accessed messages. They may delete messages only in accordance with the contract with the client. Finally, network operators and service providers are obliged to take the technical and organisational measures necessary to avoid transmission errors and unauthorised disclosure within their organisation and to third parties. Measures are only required if their effort is appropriate to the desired purpose of security. Within this framework, measures have to be adapted to the technical state-of-the-art.

A change in the German law regulating wiretapping [31] is also relevant. Private network or service operators are obliged – as Deutsche Telekom was in the past – to reveal any data on communications and the content of messages to the secret services and to support them in their surveillance of communications. For this purpose, they must provide personnel who have been checked by the services. According to Article 10, employees of a network or service provider must not reveal whether a surveillance has taken place. This is enforced by a penalty of imprisonment of up to two years.

In France, CNIL had to decide on a regional pilot for Minitel. According to Gebhardt (p 165) [32] CNIL required free non-listing of numbers as well as security by a personal password and other measures against unauthorised access to electronic postboxes.

In the UK, British Telecom has developed privacy guidelines. Its text and voice-mail systems not only offer passwords for the boxes but additional passwords for each message.

It is difficult to overlook the situation in the US. There still seems to be uncertainty as to whether e-mail falls under the wiretapping acts which are slightly different in each state (see Dutton) [33]. In company networks, surveillance of employees' mail seems to be quite usual. According to a survey by Macworld, 74% of the employers admitted that they searched electronic work-files, 42% had searched workers' e-mail, and 15% had searched voice-mail (*Communications of the ACM*, August 1993, p 9). The American E-Mail Association issued guidelines for companies to develop their own privacy rules. Legislation regulating electronic surveillance has not yet been passed.

A.2.3. Information services (videotext)

In most countries, videotext systems contain a variety of distinct services, including e-mail, transactions with computer systems for teleshopping, telebanking, etc. These will be discussed separately. This paragraph deals only with information services. Most of the EU member states have their own national videotext system. Not only do they have different names but there are also great technical differences. For example, the German Bildschirmtext system is based on personal identification and central registration of each access, while the French Minitel offers the 'kiosk system' with flat rates in order to ensure anonymity.

The group of experts on data protection of the Council of Europe sees the main problem as access to and the secondary use of data concerning the content of an individual's inquiries, and which are stored for billing purposes or commercial interests, or both [8] (p 25). In contrast to the issue of itemised billing for the telephone service, itemisation of billing-data reveals much more about the behaviour and circumstances of individuals. Because of the wide range of information and other services, these data might be put together into personal-communication or consumer profiles, which, according to the experts, are 'marketable commodities in themselves'.

The experts ask for new regulators' initiatives which should concentrate on the following issues [8] (p 25):

'i. the consent of the user prior to the installation of videotext technology in his residence;

ii. only personal data which are necessary for billing purposes or service purposes should be collected and stored;

iii. the sale or disclosure or further use of the data should only be possible with the informed consent of the user, or if authorised by a legal provision;

iv. the security and confidentiality of personal data which have been recorded should be guaranteed, for example, by means of encryption techniques, smart cards, password variations, flat rates;

v. strict conservation periods should be laid down;

vi. user rights (access, deletion and rectification) should be provided for the user, who should be informed of their existence and shown how to exercise them;

vii. the general data-protection legislation, where such exists, should provide the overall regulatory framework.'

There are several technical and organisational ways of reducing the amount of personal data collected. The German special regulation for videotext, for example, rules that only the number of the last six pages called may be stored at all. In many countries the network operator, ie, the PTT, takes over the billing for information suppliers so that the supplier does not know which information has been used by which customer.

Yet the categorisation of videotext in legal terms is not obvious. In Germany, the states described it as a kind of broadcasting and claimed regulatory competence. Videotext is blurring the boundaries between media which used to be regulated quite differently. With the further convergence and the use of multimedia systems, it will become even more difficult.

Other problems in this context concern the conflict between freedom of expression and the control of information with sexual, racial or other discriminating content. Should the service-provider exercise control over the information offered by suppliers and should it have the right to delete certain information? Or should other bodies take over this responsibility?

The German regulation for Bildschirmtext also contains elements relevant to consumer protection. As single pages can be charged with different amounts by the supplier, this has to be announced before access is provided.

A.2.4. Teleshopping

Videotext systems often include teleshopping services. But there are other technical means, in particular a combination of presenting goods on television and asking for orders by phone (see Wolsing) [34]. Some of the problems are the same as those with mail order. Therefore, the CEC has presented a draft of a directive concerning consumer protection in distant selling where distant selling includes any kind of contract where the supplier and consumer are not present at the same time but rather use a telecommunications technology. According to the appendix of this draft, this technology may be a letter or an order form, a telephone call, fax, e-mail, videophone, voice-mail system, television, etc.

The main objective of the draft directive is to harmonise legislation by the respective member states. Therefore, it mainly states general principles:

- Member states shall take the necessary measures so that the means employed for distant selling:
  - consider the privacy of the consumer and in particular correspond to Article 17 of the specific directive on data protection and privacy in telecommunications;
  - be in accordance with the principles of good faith in commercial transactions and those for the protection of minors;
  - make obvious the commercial purpose;
  - inform the consumer on any charges for the transmission of an order, if it is not obvious.
- Suppliers have to provide information about set details, such as the identity of the supplier, relevant features of the product or service, the price including additional costs for transport, etc.
- Consumers have the right to withdraw from any contract without justification within seven days after receiving the product or service; they only have to pay for the direct costs of returning the product.

In order to protect consumers paying by credit card, the directive rules that in cases where only the card number has been communicated without presentation or electronic authorisation, the owner of the card may cancel. Such cancellation results in the amount already transferred to the account of the supplier being immediately returned to the account of the consumer.

This draft has not yet passed the Council. As it tries to cover any form of distant selling, it is not very precise on data protection. It is not clear why a reference is made to Article 17 of the draft of a special privacy directive which deals with unsolicited calls, but not to Articles 4 and 16 which regulate the storage of only those data which are necessary and stipulates that no personal profiles may be set up.

As Article 16 has been eliminated, the consideration of privacy regulation in the directive on distant selling becomes even more urgent. Here we have a basic problem in the regulation of privacy in value-added services. Because different DGs and Parliamentary committees are responsible, it is always difficult to coordinate the initiatives.

A.2.5. Telebanking

The term telebanking here refers to any use of bank services by telecommunications facilities located outside the bank. The most prominent forms are automated teller machines (ATM), homebanking, in particular by the use of videotext systems, and EFTPOS (electronic funds transfer at the point of sale), where network terminals are located near the cash register of supermarkets or gas stations, in restaurants, etc.

Scanning through the literature, the following problems for data and consumer protection are mentioned (for a summary see Klein and Kubicek) [35]:

(1) Security of the system, in particular of the PIN (personal identification number): for homebanking, for example, by videotext, in many countries there is a high level of access control. In Germany, a PIN and additionally a transaction number for each transaction (TAN) is required. EFTPOS systems, however, vary widely. Credit-card systems usually do not require a password.

(2) Distribution of liability and of the burden of proof: because of the lack of security it becomes a crucial question as to who is liable if funds have been withdrawn from a consumer's account which he claims not to have authorised (see Poullet) [36]. There are different regulations concerning liability in the case of loss or theft of the card, even between commercial banks and saving banks in Germany.
(3) Personal profiles: newspapers report that credit-card organisations generate consumer profiles in order to detect misuse. There is definitely a common interest in a high degree of security. But how much data may issuers of cards store and process and for how long?
(4) Dissemination of data to third parties: in some countries issuers of cards or processing centres, or both, sell customers' data.
(5) Combination with other data: so far, payment data do not include details of the particular items that have been paid for. But with increasing use of computers in supermarkets and their itemised billing systems, it becomes technically feasible for retailers to combine the anonymous sales-related data and the personal payment-related data.
(6) The revealing of account-related data in the authorisation: if EFTPOS systems employ an online authorisation, an electronic message is sent back to the customer at the card terminal and to the retailer's printer. In Germany, if a consumer's account does not provide for the amount requested, the terminal only shows 'Transaction not possible', but a code number is printed into the retailer's journal, thus revealing that the account would have been overdrawn.
(7) Application and enforcement of data-protection rights: data-protection law gives the individual the right to know where data are stored and processed. However, EFTPOS systems are very complex. If I pay at a gas station, some data are stored there and some data are stored in the processing centre of the mineral oil company, as well as in the two banks involved. If I pay in a shop, usually a network operator is involved. Some banks do not authorise transactions themselves but employ an authorisation centre, and this means that a consumer usually does not know where to ask for information about his personal data. In addition, German banks claim that EFTPOS transactions do not involve personal data. They only include account numbers which, in their view, do not allow for identification of individuals and, therefore, do not fall under the data-protection act.
(8) Budget control: consumer organisations claim that it is becoming very difficult for consumers firstly to control their budgets, when it takes some time before funds are withdrawn from their account and, secondly, to check for misuse.

Questions of data or consumer protection, or both, in EFT have been subject to legislation in the USA, France, and Denmark (Poullet[36]; Vandenberghe [37]). The Danish Payment Cards Act of 1984 restricts the liability of card holders as well as the storage and forwarding of their personal data and introduces a consumers' ombudsman to enforce the provisions of the Act.

In 1987, the CEC issued recommendations for a European code of conduct in the field of electronic payments, in particular concerning the relationship between financial institutions, retailers or service providers and consumers [38]. Consumer organisations were disappointed that the Commission had failed to agree on a legally-binding directive (European Consumer Law Group) [39]. With regard to the content, the ECLG demanded a regulation of the card issuer's liability which is not mentioned at all in the recommendations. In addition, they asked for some 'dispute and error settlement procedures'.

While the Commission did not respond to the claims of consumer organisations, in September 1990 the Committee of Ministers of the Council of Europe adopted a recommendation on the protection of personal data used for payment and other related operations [40].

This document maintains that the Data Protection Convention of 1981 needs to be amended by more specific provisions dealing with the particular risks to privacy raised by electronic-payment systems. In order to promote the transborder data flow in payment systems, they recommend that member states integrate the principles contained in this paper in their domestic law. These principles concern the collection and storage of data, the use and the communication of these data, the rights of access of the persons concerned and the security of data:

- The providers of means of payment and the beneficiaries, for example, retailers, may collect and store personal data only in so far as it is necessary for payment and verification.
- Payment systems should be designed in such a way that in the course of payment they do not communicate personal data which are not required.
- Personal data may only be used for other purposes if the individual has been fully informed in writing and has not objected.
- The interconnection of different personal-data files is only permissible under specified conditions.
- Beneficiaries and network operators have to ensure that the individual is informed about the storage of personal data, the purposes of processing and the bodies to whom data are communicated.
- The body providing the means of payment has to ensure that the individual can exercise his right to access, in particular when several files are involved in a system of distributed data processing.
- Member states should secure respect for these principles by instituting a supervisory body.

A.2.6. Audiotext

Audiotext services were introduced later than videotext services but might be used more frequently by now in the residential areas of some countries. They raise similar problems with regard to privacy, consumer protection, and protection of children and young people. Public debate concentrates on the prices and content of most of the services offered.

Compared with traditional announcement services, audiotext services charge extremely high prices. In Germany, for example, they cost 1.50 DM per minute in most cases. Advertisements feature these prices because of an agreed code of conduct. In the US, pricing has been a hotly discussed issue, which has led to several Congressional initiatives. Prices there vary widely between service providers. Another issue is the protection of children from the pornographic content of such services ('dial-a-porno').

In California, a bill (entitled SB 679, Rosenthal) has been drafted which requires that services reserved for adults are given a special number that can be blocked individually by subscribers. Another draft requires that the first twelve seconds have to be free of charge, so that the customer can get a notion of what he will be paying for.

As there are major privacy concerns related to registering the use of these services, network operators, and in particular PTTs, usually take over billing. But as far as I know, there is no regulation that prevents service providers from registering users, for example, by calling-line identification, and that forbids the generation of communication profiles.

According to the German special privacy regulation for telecommunications, network operators may register every connection to a service provider. On request, the customer gets a list of every call giving details of the name of the service provider, the date and length of the connection and the amount charged. If he does not pay, PTT is entitled to give these data to the respective service provider.

In Germany, there is still discussion about whether audiotext is broadcasting and, therefore, has to be licensed under the media law (which is state law).

On the EU level, audiotext services have not been incorporated explicitly into the special data-protection directive. Neither is the relationship between network provider and service provider generally regulated there.

As a big market is expected for audiotext services, the Legal Advisory Board to the CEC established within the IMPACT programme has undertaken consultation with the objective of reaching a harmonised regulatory framework which allows for transborder audiotext and videotext services. They had a study carried out by the Computer Law Institute of the Free University of Amsterdam on the status quo of regulation and the topics to be included. These topics are: conditions of access, content of information, liability and security, data protection, and user protection. The most controversial issue was that of content.

It is interesting to note that the European Information Industry Association has taken the initiative and presented guidelines which acknowledge that some services need particular regulation and should be subject to additional controls. In particular, they mention 'competitions, lotteries, children's services, fundraising, sales promotions, chatlines, sex services, and other services of a political, religious or moral nature' (EIIA) [41]. For other services, the guidelines contain minimum requirements concerning price information, privacy, and content. For example, services should not induce or promote racial disharmony, contain false or out-of-date information or be unreasonably prolonged or delayed.

The Association suggests that these guidelines be made part of the contracts between network operators and service providers. They would be supervised by the national regulatory organisation responsible for audiotext and videotext. Complaints about transnational services would be handled by a new supervisory committee, the European Supervisory Committee for Audiotext and Videotext.

Notes and references

[1] Council of Europe, Convention on the Protection of Individuals with Regard to Automated Processing of Personal Data, 28 January 1981.

[2] See Rieβ, p 751 [12].

[3] CEC 1987, see also Ungerer 1989 [9].

[4] Commission Directive of 28 July 1990 on competition in the markets for telecommunications services, 90/388/EEC.OJL 192/10, 24.7.1990.

[5] Council Directive of 28 June 1990 on the establishment of the internal market for telecommunications services through the implementation of open network provision (ONP), 90/387/EEC-OJL 192, 24.7.1990

[6] Proposal for a Council Directive concerning the protection of individuals in relation to the processing of personal data (SYN 287, Com (92) 422 endg.).

[7] It is worthwhile to note at this point that the most heated debate on calling-line identification did not take place in Germany or another country known for its great concern with data protection, but in the US. There, the regulatory bodies of nearly every state have been occupied with this issue over the last three years (see *Privacy Journal*, November 1992).

[8] Council of Europe, New Technologies: a challenge to privacy protection? Study prepared by the Committee of experts on data protection under the authority of the European Committee on Legal Co-operation, Strasbourg, 1989.

[9] Herbert Ungerer, Einführung in die EG-Telekommunikationspolitik. In: Herbert Kubicek (Ed.), Daten- und Verbraucherschutz bei Telekommunikationsdienst-leistungen in der EG. Nomos, Baden-Baden, 1993, pp 23–31.

[10] Alexander Roβnagel, Vom informationellen zum kommunikativen Selbstbestimmungsrecht. In: H. Kubicek (Ed.), Telekommunikation und Gesellschaft. Kritisches Jahrbuch der Telekommunikation. C. F. Mueller, Karlsruhe 1991, pp 86–111.

[11] Eli M. Noam, Privacy in Telecommunication Services. In: Der Berliner Datenschutzbeauftragte (Ed.), Data Protection. Convenience and Rights of Telecommunications Consumers. Proceedings of a workshop, Der Berliner Datenschutzbeauftragte, Berlin, May 1992.

[12] Joachim Rieβ, Daten- und Verbraucherschutz im europäischen Telekommunikationsrecht, *Computer und Recht*, Vol. 7, 1991, pp 747–753.

[13] Council Resolution of 22 July 1993 on the review of the situation in the telecommunications sector and the need for further development in that market (93/C 213/01).

[14] Eli M. Noam, Telecom Privacy Policy Elements, *Transnational Data and Communication Report*, March 1990, pp 9–19.

[15] State of New York, Department of Public Service, Strategies for Implementing Public Service Commission Telecommunications Policy in the 1990s. New York, December 1992.

[16] W.F. Korthals Altes, Telecommunications, Itemization, and Privacy: some developments in the EC and the Netherlands, *Media Law Journal*, Vol. 2, No. 1, Spring 1993, pp 3–6.

[17] Heinzpeter Höller, Kommunikationssysteme - Normung und soziale Akzeptanz. Braunschweig. Vieweg, Wiesbaden, 1993.

[18] Commission of the European Communities, Directorate General XIII, Information Technology Security Evaluation Manual. Draft V.02, Brussels, 1992.

[19] Legal Advisory Board, Legal Questions of European Audiotex and Videotext. Meeting of 11 and 12 March 1992, File No. 92/1, Commission of the European Communities, 1992.

[20] Council of Europe, Working Party No. 9 of the Committee of Experts on Data Protection, GT/9 (New Technologies: telecommunications sector). Report on the 5th meeting, 27–29 November 1989, Strasbourg, 12 January 1990.

[21] The expert group of the Council of Europe speaks of content data, which might be misleading, because the content of the communication is not stored, only data on the conditions (the numbers involved, at what date and how long the communication lasted).

[22] Detlef Garbe, Datenspuren in digitalisierten Telekommunikationsnetzen. Ein Vergleich der Datenschutzregelungen in europäischen Ländern im Fall ISDN. Diskussionsbeiträge No. 103. Wissenschaftliches Institut für Kommunikationsdienste, Bad Honnef, December 1992.

[23] H.-P. Gebhardt, The Legal Basis of Data Protection and Data Protection in the Field of Telecommunications in Seven Countries: Switzerland, France, the Federal Republic of Germany, the United Kingdom, Sweden, the United States and Japan, *Telecommunications Journal*, Vol. 57, No. 1, 1990, pp 37–44.

[24] There are two directives with almost identical content, one for the Deutsche Bundespost Telekom: Verordnung über den Datenschutz bei Dienstleistungen der Deutschen Bundespost Telekom (Telekom-Datenschutzverordnung - TDSV) vom 24. Juni 1991, BGBl. I, S. 1390ff., and one for private telecommunications enterprises: Verordnung über den Datenschutz für Unternehmen, die Telekommunikationsdienstleistungen erbringen (Teledienstunternehmen-Datenschutzverordnung - UDSV) vom 28.12. 1991, BGBl. T, S. 2337ff.

[25] James E. Katz, Caller-ID, privacy and social processes, *Telecommunications Policy*, Vol. 14, 1990, pp 372– 411.

[26] Marc Rotenberg, Communications Privacy: implications for network design, *Communications of the ACM*, Vol. 36, No. 8, August 1993, pp 61–68.

[27] Paul Slaa, Telecommunications and Privacy. The 'caller ID controversy' as an example. Paper presented at the session on Privacy and New Media of the Mediaforum, Berlin, 31 August, 1989.

[28] Paul Shultz, Caller ID, ANI and Privacy. A review of the major issues affecting number identification technologies. Telecommunication Reports Inc., Report Series No. 4, New York, 1990.

[29] Heinzpeter Höller and Herbert Kubicek, Gestaltung digitaler Nebenstellenanlagen. Kommunikat ionsökologische Probleme und technische Anforderungen an Leistungsmerkmale. In: F.L. Holl and R. Schlag, Digitale Telefonanlagen. Die neue Welt der Kommunikation. Bund Verlag, Cologne, 1989, pp 155–174.

[30] Herbert Kubicek (Ed.), Daten- und Verbraucherschutz bei Telekommunikationsdienstleistungen in der EG. Nomos, Baden-Baden, 1993.

[31] Gesetz zur Beschränkung des Brief-, Post- und Fernmeldegeheimnisses vom 13. August 1968, geändert durch Artikel 4 des Poststrukturgesetzes vom 8. Juni 1989.

[32] H.-P. Gebhardt, Rechtsgrundlagen des Datenschutzes sowie Datenschutz im Fernmeldewesen der Länder Schweiz, Frankreich, Bundesrepublik Deutschland, Groβbritannien, Schweden, USA und Japan, *Archiv für das Post- und Fernmeldewesen*, Vol. 41, 1989, pp 163–169.

[33] William H. Dutton, The Social Impact of Emerging Telephone Services, *Telecommunications Policy*, Vol. 16, 1992, pp 377–387.

[34] Theo Wolsing, Verbraucher- und datenschutzrechtliche Probleme des Teleshopping. In: Herbert Kubicek (Ed.), Daten- und Verbraucherschutz bei Telekommunikations-dienstleistungen in der EG. Nomos, Baden-Baden, 1993, pp 126–134.

[35] Stephan Klein and Herbert Kubicek, Fields of Conflict in Card Based Payment Systems. What about data protection - what about consumer protection, à *la Card Euro-Journal*, 2/1991, pp 22–51.

[36] Y. Poullet, E.F.T. und Datenschutz, *Datenschutz und Datensicherung*, No. 2, 1988, pp 64–73.

[37] G. Vandenberghe, Rechtliche Aspekte verbraucherorientierter Telebanking- und Teleshopping-Dienste in der Europäischen Gemeinschaft. In: J. Scherer (Ed.), Nationale und europäische Perspektiven der Telekommunikation. Nomos, Baden-Baden, 1987, pp 149–157.

[38] Recommendation 87/598, December 8, 1987, OJL 365, 24.12.1987, p 72.

[39] European Consumer Law Group, New Electronic Means of Payment. Arguments in Favour of Consumer Protection with Regard to the Contractual Aspects, *Bureau Europeen des Unions de Consommateurs, Legal Supplement*, October/November 1988.

[40] Recommendation No. R (90)19 of the Committee of Ministers to Member States on the Protection of Personal Data Used for Payment and Other Related Operations, adopted by the Committee of Ministers on 13 September 1990.

[41] European Information Industry Association, Guidelines for Cross Border Control of Audiotex and Videotext Services in Europe. Luxembourg, July 1993.

Telecommunication: Exploring Competition
H. Williams and M. Borman (Eds.)
IOS Press, 1994

Chapter 8

Reforming Greek Telecommunications: the complexities of change in a Southern European country

Yannis D. Caloghirou
National Technical University of Athens, Greece

Abstract: Following a long period of network expansion and institutional stagnation, Greek telecommunications entered a phase of modernisation, restructuring and reform in the 1990s. During 1990–1993, privatisation was the main policy instrument selected by the conservative government to promote the rapid reorganisation of the sector. This was never implemented as, in addition to other public-policy disputes, it produced a political crisis that led to the fall of the government and the breakdown of its telecommunications policies. This paper summarises important developments in this period and provides tentative conclusions on the policy-making process in Greece in the light of telecommunications developments.

1. Introduction and summary

Greek telecommunications, after a decade of waning network expansion based on the old analogue-monopoly paradigm [1], entered a turbulent period of radical restructuring in the 1990s. This period has now reached a critical stage, highlighted by political uncertainties and a shift of power from the conservatives back to the socialists in the autumn 1993 general elections. With telecommunications unanimously acknowledged as a central feature of the recent political turmoil, any study of telecommunications-specific developments necessarily has to involve an assessment of a broader spectrum of economic and social reform measures.

In Greece, as in other countries, attempts to match technological with institutional change have encountered a unique set of economic, social and political conditions, which defy any mechanistic transfer of reform models implemented in other world regions [2]. In particular, the Greek telecommunications experience supports the case for innovative, flexible policy-making in response to evolving local conditions, based on a critical assessment of the experience of other countries. In this respect, it is hoped than this paper, in addition to its Greek focus, may prove useful to policy analysts in other countries faced with a similar policy agenda.

The need for restructuring Greek telecommunications has been caused by a number of factors: first, by international developments, particularly those taking place at the European Union level; second, by the last conservative government's launching of an economic programme of privatisations mainly as a means to alleviate burdening public-sector deficits; last but not least, by the decaying state of the infrastructure itself, which clearly inhibits economic development in all sectors, including those critical for the recovery of the Greek economy (tourism, shipping, etc).

From the first instance, it was clear that telecommunications would be raised to a principal issue, to be resolved in the country's central political arena. Indeed, the scope of necessary changes, as well as the significance of their institutional, economic and social implications demanded that decisions should be taken at the highest possible level, by those democratically elected, through the implementation of dialogue and a consensus-building process with all involved parties.

On the other hand, the sheer complexity of the overdue changes, as well as the strong organised economic interests that are de facto linked to alternative scenarios, favoured a 'traditional' approach to the problem. In the Greek case 'traditional' refers to political manoeuvring or clashes, or both, behind the scenes, rather than open public-policy debates. This sort of arrangement is effective in providing direct access to public markets, through the exploitation of political contacts. On the other hand, it tends to reproduce rather than reform existing economic structures. In addition, where equal opposing interests collide, it may lead to a stalemate, expressed through public-policy inertia. Finally, when exposed to the public eye, it increasingly leads to the public revelation of a 'scandal'.

In the light of the above arguments and the current debate, this paper examines the positions and actions of principal actors involved, including: the government; the European Commission; major political parties; foreign and local-based telecommunications-equipment manufacturers; business users; residential users; financial/stock market consultants; future investors; and trade unions. The political, economic and social implications of the proposed changes are analysed and discussed.

Moreover, the broader consequences of the telecommunications reform with respect to the conduct of public policy are assessed. In particular, the willingness and ability of the last (or indeed any) government in Greece to promote its policies through the organisation and running of a well-structured public debate is being put under test. In the case of telecommunications, a mode of policy practice in which there is a non-transparent compromise between conflicting vested interests, with overall policy inertia prevailing, may be deeply rooted.

Expanding on the above, the paper concludes that the political and social context of the process of change introduced in the Greek telecommunications regime is determined by the balance between the structural inertia and reaction to change which governs Greece's political and social life, as well as by the pressing requirements for reform associated with EU membership. The imperfect function of the market and the absence of indigenous, strong private actors in the Greek economy, emphasise the role of political elements (political parties, the press) often acting as agents of vested (economic) interests concealed behind a prevalent statism. In such conditions, the challenge to the state is to use its dominant position to overcome a policy of myopic compromises and install a framework for dynamic market development in the telecommunications sector.

In order to investigate the above hypothesis through an assessment and analysis of the present Greek telecommunications situation, this paper adopts the following structure:
Section 2. summarises all important developments that have been initiated by the Greek conservative governments of the period April 1990–October 1993.
Section 3. discusses how and why these developments have led to an escalating political crisis, culminating in the fall of the conservative (New Democracy) government and the return of the socialist party (PASOK) to power, after the October 1993 elections.
Section 4. provides some basic conclusions concerning the policy-making process in Greece in the light of telecommunications developments.

2. The drive to restructuring

Greece has embarked, since 1990, on an ambitious restructuring of its telecommunications system. The overall goal of Greek efforts has been to 'close the gap' between Greece and other EU states so that the country can best be integrated into the emerging European (and global) markets.

Table 1. Main indicators of Greek telecommunications and OTE performance

	OTE (1990)	OECD Average (1990)	OTE (1992–93)
Main lines per 100 inhabitants	38.94	42.54	45
Digitalisation ratio for local lines	0.5	36.4	14
Revenue per main line, US$	318.94	1029.29	315
Percentage of revenue from PSTN	85.49	72.66	
Profit per total revenue (%)	21.8	12.7	31.6
Revenue per capita, US$	124.2	437.8	142.5
Services market as % of GDP	1.85	2.28	
Investment per main line, US$	87.71	240.08	
Main lines per employee	141	145	173
Revenue per employee, US$	44800	149700	54800
Faults per 100 main lines	55	10.4 (France)	
Waiting time for telephone installation (1990), months	81	3	17

Source: OECD [3], Technical Chamber of Greece [4], OTE [5]
Note: 1992–93 calculations are provisional

Table 1 provides a summary of the current state of the Greek telecommunications environment, focusing particularly on the performance of the monopolist supplier, OTE. Despite the relatively high degree of telephone main-line penetration, the infrastructure remains predominately a patchwork of analogue technologies. As a result, service quality offered by OTE continues to be poor. Moreover, institutional stagnation has left advanced telematic-service markets underdeveloped. Greek tariffs, despite restructuring during the last two years, continue to be heavily biased in favour of local calls (mostly still untimed and offered at 55% of cost) and are consequently also candidates for restructuring.

Given the above situation, the development of Greek telecommunications involves three inter-related challenges:

(i) Accelerated investment to modernise the infrastructure.

(ii) Institutional changes that will harmonise Greek regulation with the European Union regulatory model.

(iii) Deep organisational reforms within OTE itself, including management restructuring and changes in its ownership status.

So far, restructuring has proceeded in the following areas:

2.1. *The procurement of digital switches saga* [6]

Although the need to go digital was recognised as early as 1977, it was only in 1988 that the first pilot contract was awarded. The major reason for this twelve-year delay was the inability of various Greek governments to implement a selection process that would determine the switch suppliers as well as the procurement conditions (such as, price, establishment of domestic production facilities with local value added, etc.). Behind the stagnation lies decade-old clashes between major suppliers for the control of the Greek market.

In 1988, two small pilot contracts were awarded: one to Intracom, a Greek firm manufacturing Ericsson switches; and one to Siemens Hellas, a subsidiary of Siemens. These were followed by a more substantial contract, awarded in early 1990 by a short-lived all-party coalition government, followed by an extension granted in 1991. Thus, it seemed that the two suppliers became established in the Greek market, although it was less clear whether there was genuine competition between them.

Other contenders however, especially Alcatel, AT&T and Northern Telecom have not ceased to fight for market entry. The question of a third, or even a fourth supplier in a market as small as Greece was for all practical purposes left open. In a recent (1993) open bidding for an additional one million lines, all prospective suppliers participated, with 'outsiders' apparently offering knock-down prices to achieve a market breakthrough. No decision has been taken yet, especially in the light of OTE's privatisation process, with the final report on the suppliers' selection prepared at the end of July 1993 by OTE's procurement division. As discussed in section 3, the disagreement on this matter between OTE and the Ministry of National Economy led to a major political crisis.

2.2. *The telecommunications sector strategic-study initiative*

The modernisation of the infrastructure, initiated in 1989 [7], has already contributed to the improvement of the situation regarding basic-service provision. However, progress in this field alone is not yet sufficient, especially in view of the major findings of an independent study, conducted in 1990–92, with the aim of formulating a short-term 'crash programme' for OTE, as well as a ten-year telecommunications strategy and a 'business plan'.

The study was awarded to Coopers and Lybrand, and was financed by the European Commission (US$2.4 million). The idea that a foreign firm, rather than OTE itself, should lay down the direction lines for the sector's future was met with hostility by the Greek telecommunications establishment. The study's major findings point to three areas of reform [8]. First, a drastic re-organisation of OTE's management, effectively realising a true corporatisation of OTE, with the option of partial privatisation left open. Second, an intensification of investment that will enable Greek telecommunications to become equal or superior to other European ones by 1993. Third, regulatory reform, including a restructuring of tariffs to bring prices in line with costs.

The importance of this study stems from the fact that it was used as a 'restructuring bible' by the last conservative government, which aligned its policies to-the-end with the study's major recommendations. Moreover, in the context of the study, the 'crash programme' (200 million ECU, 50 per cent funded by the Commission) has been initiated, to improve the network where major demand exists and, as its proponents put it, to 'stop the ship sinking'.

2.3. *Cellular telephony*

In mid-1993 a cellular service, based on the GSM digital pan-European standard, was launched in Athens. The service is offered by two competing providers, STET Hellas (STET 95%, Interamerican (a Greek insurance company) 5%) and Panaphon (Vodaphone 45%, France Télécom 35%, Intracom 10%, Data Bank 10% – Data Bank is owned by the Greek Lambrakis Press Group and Intrasoft, of the Greek Intracom Group). Both operators bid US$163.40 per user for obtaining a licence for a total market of 160,000 potential users throughout Greece by 1997. The number of subscribers, according to some optimistic projections, could reach 500,000 by the year 2005. According to Panafon, the total investment for that market will reach 130 billion drs (US$567 million in 1993 prices).

This result has been achieved following a complex and somewhat controversial process. The reform was initiated in 1991, when the newly-elected conservative government decided to hold a competitive bidding, with OTE participating by 30% in both winning consortia. This term, which was a compromise between OTE's desire to have a 100% monopoly and a truly-open competition, was subsequently annulled. Thus, OTE was left out of the bidding process and was by law (July 1992) excluded from all cellular activities. The final bid was awarded shortly afterwards (August 1993). OTE's management establishment attempted to delay the process of interconnection arrangements between its terrestrial network and the cellular operators, but the dispute was resolved leading to an earlier than scheduled commencement of service. This dispute was only part of broader context of tense conservative government-OTE relations, since the government's decision to privatise OTE.

2.4. *OTE's partial-privatisation strategy*

The privatisation of OTE was the dominant issue in the Greek telecommunications scene throughout 1992–93. The last government was determined to transfer 35 per cent of OTE plus the management responsibilities (for a decade) to a foreign strategic investor, sell seven per cent through international stock exchanges and another seven per cent through the Athens Stock Exchange. The whole process was to be completed by the end of 1993, with September 1993 a scheduled decision deadline.

According to its supporters, OTE's privatisation is the only policy option available in order for it to become competitive in an increasingly internationalised market place. This is to be achieved through an 'internal shock' that will revolutionise its management structure, with the strategic investor transferring its managerial and technical expertise, as well as financing the large investment requirements of the organisation (5.3–7.5 billion ECU for the next decade). According to the ex-Minister of National Economy, who was masterminding the whole privatisation process, the necessary investment funds need not be drawn out of EU resources (which can then be directed to other public-investment priorities) and cannot be covered by the Greek state. Without a commercial approach to management practices, OTE's profitability will be increasingly eroded by international competition, especially with regard to its international traffic revenue.

From a shorter-term perspective, in addition to solving the longer-term OTE problems, privatisation revenues will contribute to the immediate financing of the budget deficit and to the boosting of the Athens Stock Exchange operations. In the latter case, the government expects to trigger a phenomenon of 'popular capitalism', similar to that which occurred in Thatcher's Britain.

The above plans encountered strong opposition from a broad anti-privatisation camp, which includes trade unions, all other political parties (plus leading ruling-party figures), the defence establishment and the present OTE management. The Confederation of Greek Industries (representing large users as well as important domestic suppliers) has assumed a neutral position, emphasising the need for regulatory restructuring.

Arguments against the privatisation centre around the need to sustain national management control in the sector. This is thought to be necessary for reasons of national development policy (leading technological role for OTE in the industrial structure), defence policy and social policy (universal service, employment). In addition, privatisation opponents claim that OTE (which is a net contributor to state funds) will be sold at an unacceptably low price.

In the first stages of the privatisation process (February 1993) 13 candidates were short-listed for the strategic-management package. They included five European operators, six North American carriers and two Asian operators. Subsequently, it was reported by Reuters that six firms among them were the most involved (France Télécom, GTE, NTT, STET, Telefonica, Korea Telecom), although this shorter list was far from binding. The total OTE evaluation was subject to intense speculation. An approximate value could be around US$5 billion, with $1.9 billion expected from the strategic investor and an additional $0.7 billion raised through the stock markets.

With negotiations in progress, it was expected that three factors would be decisive for a successful conclusion by September 1993.

First, the ability of the Greek government to offer a promising regulatory package to the winning consortium, hence beating international competition from privatisations taking place elsewhere (for example, Eastern Europe). This is a direct consequence of the fact that the regulatory regime was not finalised prior to the initiation of the privatisation, but is being negotiated in parallel. The granting of strong (private) monopoly rights, however, may be contrary to the interests of Greek users, or other foreign investors.

Second, the extent to which 'strategic-investor' candidates will form consortia that will divide the risk between them and facilitate the privatisation's financing. The inclusion in consortia of key Greek actors – with experience in the Greek market – may be a wise tactical move that will enable 'smoother' relations between the new managers and the state.

Third, the ability of the government to diffuse some of the domestic opposition to its plans. A national consensus, as well as convincing solutions to key problems (as for example, national defence-communications control) were required. Again, the participation of some Greek actors, as well as the specification of the rights of the yet-unformed national regulatory committee, may be important steps.

Following the government's failure to meet the above requirement, the whole process has been frozen since mid-September 1993, when the premature elections were announced. The Socialist Party's policy is basically against privatisation, as discussed in section 3.

2.5. Market liberalisation

Greek telecommunications have for decades suffered the effects of political intervention. Privatisation is supposed to disentangle the sector from the state altogether but in itself does not guarantee the development of competition. In fact, if not supplemented by a regulatory review introducing competition, it unavoidably leads to the substitution of a state monopoly by a private one.

The conservative government's telecommunications policy has been criticised as a policy example where the choices of privatisation and liberalisation contradict each other, chiefly because they are being promoted simultaneously.

Greece has had, until July 1992, a regulatory environment of typical monopolistic arrangements, with OTE being a wholly state-owned enterprise and not a Ministry department. In addition, regarding its European regulatory bounds, Greece has been exempted from two important obligations, on account of its 'developing' nature in the sector. First, OTE was granted permission to sustain its monopoly in data networks (in addition to the telephony and basic-infrastructure monopoly) until 1996. Second, telecommunications-equipment procurement was exempted from the regulation of obligatory open tendering, also until 1998. Moreover, Greece was awarded a basic-service (telephony) exemption by the Commission until the year 2003.

In July 1992, the Greek government, having to comply with other European Community regulatory obligations, passed a new Telecommunications Law in the parliament [9]. The law, which was designed after the EC regulatory model, instituted an independent regulatory authority and imported the main regulatory concepts regarding the development of competition, the licensing of operators, and principles of open-network provision. However, in order to be effective, the new law required the issuing of certain administrative decrees. This had not occurred by May 1993, whereas in the interim period, negotiations with the prospective OTE 'strategic managers' had already commenced.

The delay in implementing liberalisation suggests that the conservative government had incorporated in the negotiations for the transfer of OTE's strategic management, a number of sensitive regulatory issues, which would determine the profit margins of future owners. In other words, it was clear that the previous government's choice had been a strong private monopoly which would undertake the modernisation process, rather than competition. In this respect, the basic-service exemption became influential in the negotiations and supported the case for a higher price.

3. The restructuring crisis

Following the last government's decision to proceed with its plans for the partial privatisation of OTE, all attention shifted to the political means for this to be achieved.

The method preferred by the ex-Minister of National Economy – who was responsible for all Greek privatisation plans – was characterised by decisive, fast moves designed to pre-empt criticism. Thus, in the telecommunications case it took the Ministry just nine months to complete its plans and present its law on the new legal structure of OTE to the parliament – not to mention the Telecommunications Law already passed in July 1992 concerning regulatory matters imposed by the European Commission on Greece – an amazing record by Greek standards. Ultimately however, the advocates of radical change would have to meet all political and economic opposition in order to legitimise their plans. This process was designed to run in two parallel stages: first, in the parliament, where political approval would be sought; and second, in critical unofficial negotiations with all 'highly interested' parties around OTE, in a position to influence the imminent political decision. This second stage of course was, as discussed, already activated throughout the nine-month period of privatisation developments. Nevertheless, it is normal practice that in this sort of business negotiation, the crucial decisions are the final ones.

In order to understand the conflicting character of the whole reform process and the nature of the associated public debate, it would be necessary to present the positions and actions of the main actors involved.

First, the main agent of the proposed change, the Ministry of National Economy, made an attempt to enrich the whole argument in favour of the partial privatisation case. This was done by putting emphasis on other aspects – apart from the obvious, ie, contributing to facing the acute problem of the public deficit – such as the restructuring of the Greek telecommunications system by attracting direct foreign investment and foreign management expertise in order to successfully meet the new challenges in the field [1].

Second, OTE's strong trade union, backed by the General Worker's Confederation firmly opposed any attempt to transfer even one share to private shareholders, let alone to transfer the organisation's management control [10].

Third, the foreign and local-based telecommunications-equipment manufacturers had a strong interest on OTE's future ownership status and management control and thus, procurement regime. In this framework, incumbents in OTE's switches market were interested in keeping things as they were, while potential new entrants were interested in who is going to control OTE's management functions.

Fourth, business users in general, although they are always complaining about the poor performance of the Greek telecommunications system, were not very active in promoting or opposing the government's telecommunications policy. Nevertheless, the Confederation of

Greek Industries (SEB), representing both OTE's major business users as well as its domestic equipment suppliers, expressed a criticism of the government's policy from a radically different perspective.

SEB seemed to be against the substitution of a state-owned monopoly by a private monopoly and stressed the need for the institution of a liberal regulatory regime that would realise the benefits of competition in the Greek market [4]. To promote such a regime, SEB supported the harmonisation of the Greek regulatory framework with that instituted by the Commission. On the other hand, SEB has also been careful not to upset its telecommunications-equipment supplier members, favouring planning agreements for the benefit of domestic firms active in public procurement. In that respect, SEB also agrees with the relevant exemptions granted to Greece and argues for their implementation.

Fifth, residential users in Greece are not organised as a pressure group and traditionally they have been accustomed to long waiting lists and to poor-quality services in a producer-dominated market. Although, a user group has been established, its membership is still thin and therefore, despite an attempt to participate in the public dialogue, its real weight in the balance of power concerning telecommunications matters is negligible. Certainly, the main concern of residential users refers to the expected rise in tariff levels for residential subscribers. Usually, this concern is expressed in terms of pressure through the press and the political parties' antagonism.

Sixth, financial/stock market consultants and future investors have been very much in favour of the government's decision to sell part of OTE's shares through the stock-market mechanism, because they see this process as a means of revitalising the Athens stock market.

Seventh, the military establishment, as well as OTE's cadres, opposed the whole privatisation plan adducing serious reasons of national security, indigenous-development strategy and universal public-service provision.

Eighth, the European Commission remained neutral on the privatisation issue but it exerted considerable pressure on the adoption of the EU directives.

Finally, all the opposition parties, as well as some leading figures of the governing conservative party, were opposing the government's plan. The foundation of a new party by the government's ex-foreign minister increased the pressure on the government's stability, because the telecommunications policy was one of the two main areas of criticism which the new party promoted against the government.

Given the above balance of power, the new OTE law instituting privatisation, was presented in the summer session of the parliament (with the participation of the one-third of the total number of MPs) [11]. There, the government was forced to accept selective amendments put forward by dissenting prominent conservative MPs before the law was eventually passed.

Most amendments to the original text aimed at achieving a more even-balance between the general manager, representing the foreign strategic investor interests, and the board of directors, where the state held a majority. When finally a compromise was reached, it was not clear to what extent this had been achieved, with both government sides claiming victory.

Thus, the first phase of the government's partial privatisation plan was completed amid controversy and confusion. The second phase included the process of nomination of the strategic investor and the finalisation of an agreement between the Greek government and the latter. Finally, the third phase referred to the issue of shares through the stock market.

However, even before the beginning of the second stage, the main opposition party (socialists) sent a letter to the strategic investor candidates warning them about its intention to scrap the law. This increased the tension and uncertainty and undermined further, the negotiating position of the government. At that stage, although the elections had not yet been declared, it was clear that the government could do very little towards striking an acceptable deal.

In fact, the whole process was never completed. In the middle of September 1993, the government finally decided not to award a big contract of digital switches (one million lines) to the established OTE suppliers, despite the organisation's decision following the bid's outcome. The government's choice was made on the grounds that the final decision should be taken by the strategic investor and was most likely a last hopeless attempt to accelerate developments.

It was the day after this decision was made that the government – which was backed by a marginal majority of one vote – fell, due to one of its MPs withdrawing his support. The government, through its affiliated press, linked its fall to the interests of established switch suppliers. The defecting MP included telecommunications policy in the reasons for his decision, without however referring to procurement matters.

OTE's privatisation policy became one of the main issues of the electoral campaign. On the one side, the government presented the view that if the plan was to be cancelled, there would be an additional burden on the public deficit. One the other side, the socialist party leader made clear that his party was going to reverse this policy, by maintaining state control over OTE's activities, although it would consider the case of selling shares (up to a maximum of 49%) through the stock market mechanism. This marked a significant turning point in the party's earlier position, which was 100 per cent state ownership.

Following the election results, PASOK returned to power. One of the first decisions of the new socialist government was to declare that the recent law on the new legal structure of OTE, which was part of the privatisation plan of the former conservative government, will be scrapped. However, at the same time it provided no sign on what would constitute the new telecommunications-policy agenda. Nevertheless, a less-bold privatisation is likely. The new Greek government could maintain control over OTE and keep the management of the organisation, while at the same time issuing shares (for example, 25%) in the Athens stock market in order to finance OTE's huge, by Greek standards, modernisation programme. This is a solution which has also been supported by prominent conservative MPs.

4. Telecommunications change and policy inertia: some tentative conclusions

The presentation and policy analysis of the Greek case leads inevitably to some basic conclusions concerning the policy-making process in Greece in relation to the country's industrial, social and institutional structure.

1. It appears that one can extend Levitt's concept on 'marketing myopia' [12] to the Greek public debate on telecommunications. Indeed, some of the main actors involved seem unable to 'define' their own industry and its new competitive environment correctly. In particular, OTE's bureaucracy, trade union activists, as well as most politicians from all parties, cannot grasp the meaning of change in the new telecommunications environment. They only see the technological aspects of development and grossly underestimate the managerial, customer-oriented elements, as well as the global dimensions of change. In that respect, the Greek actors lag behind even 'commercially unsophisticated' TOs of more advanced countries.
2. The absence of a commercially-oriented approach has dominated both sides of the privatisation debate. In particular, advocates of privatisation supported the substitution of a state monopoly by a private one, claiming that foreign monopolist experience can fill the gap created by the absence of market mechanisms. This is hardly a 'neo-liberal' approach, as criticised by its opponents.
3. A permanent feature of the Greek economy is the weak restructuring capacity of the state which was already observed in the 1980's when it failed to promote the industrial restructuring process in some traditional industrial sectors despite the fact that it had a heavy

involvement [6]. The telecommunications case verifies the above tradition. The state's inflexible approach to OTE's privatisation failed to reach a consensus between all forces that would favour modernisation. It is characteristic that the economic interests linked to OTE's procurement chain were not incorporated in the 'national camp', despite having received support from the state in the past.

4. The failure of OTE's privatisation, as promoted, raises the question whether a more evolutionary approach to drastic changes better suits an environment with strong institutional inertia. The 'shock-therapy' model may be less pragmatic than a cautious restructuring of pro-modernisation forces. The evolutionary approach could include a two-stage strategy:

- The first stage would deal with the internal restructuring and market reorientation of OTE's activities and the issuing through the stock exchange mechanism of 25% of the organisation's shares. This would eventually lead to a TO owned by the state but at an arm's length from the government and protected from all political or clientist interventions and in a position to face a more competitive environment.
- The second stage could include the formation of strategic alliances of OTE with other major European TOs, leading to mixed ownership and management schemes.

5. It is widely accepted that one of the most important factors affecting the success of any policy decision is timing. Thus, even a right decision made and implemented at the wrong time may ruin the chances of its success. The privatisation-policy initiative in Greece was adopted at a time when the conservative government had lost its credibility and wider consensus and was deeply divided on several political issues. Thus, the dispute over telecommunications policy and telecommunications equipment public-procurement policy was interwoven with other major public-policy conflicts, such as those concerning foreign policy. The result was a political crisis that led to the fall of the government and the breakdown of its telecommunications-policy initiative

Notes and references:

[1] Y. Caloghirou and Th. Darmaros, Internationalisation of Telecommunication Service Provision and the Greek Privatisation Debate. International Telecommunications Society (ITS) European Conference, The Race to European Eminence, Stenungsbaden Yacht Club, Sweden, June 20–22, 1993.

[2] N. Constantelou, Liberalising telecommunications markets: political externalities in the Greek case, *Telecommunications Policy*, Vol. 17, No. 6, August 1993, pp 431–445.

[3] OECD, The 1992/3 communications outlook. OECD, Paris, 1993.

[4] Technical Chamber of Greece, OTE – modern telecommunications: a critical review of the government's proposals for OTE's privatisation and their implications. Conference proceedings (in Greek), Athens, 3–4 March 1993.

[5] OTE, Telecommunication statistics in Greece (1982–1991). Athens, 1992.

[6] Y. Caloghirou, The interlocking between the purchasing power of the state and industrial activity: the case of Greece. Ph.D. thesis (in Greek), National Technical University of Athens, 1991.

[7] OTE, Five-year telecommunication development plan (1989–1993). Athens, 1989.

[8] Coopers and Lybrand, Greek telecommunications study, 1992.

[9] Greek Government Official Journal, Law 2167/1993 for OTE's new legal structure. Athens, 1993.

[10] Institute of Labour Information Bulletin, Various articles on OTE, issues 30–31, Athens, August/September 1993.

[11] Greek Parliament Proceedings, Debates on OTE's new legal structure, sessions of 11–13 August, 1993.

[12] Th. Levitt, The Marketing Imagination. Free Press, New York, 1986.

Telecommunication: Exploring Competition
H. Williams and M. Borman (Eds.)
IOS Press, 1994

Chapter 9

Neo-liberal regulation and the smaller and peripheral economies: an institutionalist perspective

Paschal Preston
COMTEC, Dublin City University, Ireland

Abstract: This paper examines the extent to which the neo-liberal model and approaches to telecommunications regulation may be regarded as 'universal' in their reach and applicability. In particular, it examines how far the current fashion for neo-liberal regulatory regimes is valid and appropriate in the case of the less-developed regions and smaller economies or networks in Europe. The paper proposes a particular institutional political economy for the analysis of current telecommunications-policy issues in these regions. It finds that current European Union and national telecommunications policies fail to take adequate account of the specific characteristics of the telecommunications industry, especially in the case of the less-developed regions.

1. Introduction

1.1. The aims and scope of this paper

This paper seeks to address and examine the extent to which the neo-liberal model and approaches to telecommunications regulation may be regarded as 'universal' in their reach or applicability. In particular, it seeks to examine whether the neo-liberal model is equally applicable in small and large economies, in the advanced as well as the less developed regions of Europe.

More generally, this paper examines the extent to which there are particular socio-economic, institutional and geographical 'limits' and barriers to the adoption and application of neo-liberal principles in contemporary Europe. It seeks to explore and develop these issues via (a) an historical analysis of theoretical and policy debates and (b) an empirical materialist analysis focused on the contemporary case of Ireland. It goes on to examine some aspects of national and EU telecommunications-policy developments, with particular reference to the small countries and peripheral regions (SCPRs) of Europe, and to identify and address some specific telecommunications policy and research issues related to the smaller and less developed economies in contemporary Europe.

The paper points to some specific problems and possible limits to the applicability of the neo-liberal model and advances a particular institutionalist perspective on telecommunications regulation in small countries such as Ireland.

1.2. Motivation and justification for this paper

Like many items of communications research and writing, my choice of topic for this conference paper was prompted by a number of concrete developments in the telecommunications-policy arena. In particular, this paper was prompted by a desire to develop a deeper understanding of two important recent developments and the linkages between them

1) The first concerned an attempt to develop an understanding of the nature and sources of one of the most pressing and prominent current policy issues in Ireland: the very large increases in local telephone charges (May 1993). Closely bound-up with the headline-grabbing issue of tariffs, there are more fundamental regulatory issues to do with the future funding, organisation and structure of telecommunications in Ireland which need to be understood.
2) The growing popularity and increasingly 'political' reach and impact of neo-liberal telecommunications policies based on the assumed virtues or benefits of competition, or both, liberalisation, 'de-regulation' and privatisation.

The tenets and assumptions of the neo-liberal approach to telecommunications regulation (and especially the stress on competition) have become hegemonic in recent years. They inform and underpin the vast bulk of the expanding body of research reports, trade-press and mass-media coverage and national and EU policy debates concerning telecommunications since the late 1980s. Indeed, the tenets of the neo-liberal approach have become so fashionable and pervasive, that the benefits and value of this approach to telecommunications are now generally assumed and asserted rather than considered and evaluated in the light of concrete circumstances in different socio-economic contexts. In most of the literature related to this area, neither the theoretical assumptions nor the empirical foundations underlying the neo-liberal approach are rarely challenged or questioned.

Thus, this paper starts from the view that the dogmatism and dominance of any one particular set of political-economic ideas and assumptions is something that should be opened up to critical scrutiny. The paper seeks to raise questions and provoke discussion around a number of taken-for-granted positions underpinning contemporary approaches to telecommunications policy-making in contemporary Europe.

This unfashionable and critical stance may be justified on the grounds that it is important to recognise that effective telecommunications policy-making in the 1990s is a highly complex and challenging process. The stakes involved in telecommunications policy-making in Europe today are highly significant for many fields of social affairs as well as economic activities. Inappropriate policy choices and decisions in the telecommunications field could have major repercussions for many years to come.

2. Universal 'laws' and political economy: a retrospect

2.1. The role and nature of political economy in 19th century Ireland

Political economy, was the most powerful secular discourse of the nineteenth century and it was one which claimed scientific neutrality. It attempted to distance itself from moral questions as Marx [1] and other critics argued. It also claimed to be founded on natural (and hence universal) laws. In the nineteenth century Irish context, it has been characterised as 'covertly and overtly ideological, an establishment discipline geared to supporting the status quo'. Its abstract generalisations involved a denial of Irish difference. Essentially it argued that the same political economy (sets of institutions, regulations and 'structures') appropriate to England applied with equal force in Ireland (Whelan, p 35) [2].

The Irish famine was the last disaster of its kind in Western Europe. Its scale and duration was such that Ireland was irrevocably changed in its political, social and economic structures and attitudes (Deane et al) [3]. In the years of extreme socio-economic distress following the famines of the 1840s, and the subsequent political struggle and turmoil concerning the 'land question' in Ireland, the universal and other claims of the alleged 'laws' of political economy were strongly challenged. Influential Irish nationalist commentators, such as John Mitchel, attacked their influence: 'it was a perfect folly to be dancing a will-o-the-wisp dance after the abstract principles of political economy' (Whelan, p 35 [2]; Boylan and Foley [4]; Deane et al [3]).

But within the academic realm, one key figure in the challenge to the 'universal' reach of the abstract principles of conventional political economy was John Elliott Cairnes.

2.1.1. John Elliott Cairnes

Cairnes was born in Castlebellingham, County Louth in December 1823. His family operated a brewery. He studied at Trinity. In 1856–6 he held the Whately Chair of Political Economy at Trinity and in 1859 he was appointed to the Chair of Jurisprudence and Political Economy at Queens College, Galway, which he held until 1870. In 1866, he was appointed to the professorship of political economy at University College, London. He resigned from the latter position in 1872 due to ill health.

Cairnes provided an intellectual challenge to some of the dogmatic orthodoxies of his day concerning the nature and implications of political economy for practical policy-making. He was a close friend and major influence on J S Mill (see Murphy [5]; Deane et al [3]; Boylan and Foley [4]; Whelan [2]). He was a major influence on Mill's revisions to his earlier writings on the principles of political economy.

Cairnes' work has also proved to be an important influence on later twentieth century 'heretics', especially institutionalists such as Veblen and Keynes. Veblen commended Cairnes for attacking (Bastiat's and others') quasi-theological belief in the doctrine of harmony of interests which had provided a metaphysical underpinning for the economic policy of laissez faire; Veblen commended Cairnes for 'making the foundations of economics more scientific, though he realised that the opportunity cost of this increased scientificity was a concomitant decrease in metaphysical charm' (Boylan and Foley, in Murphy, pp 111–112) [5].

In 1926, Keynes declared that Cairnes 'was perhaps the first orthodox economist to deliver a frontal attack upon laissez faire in general' (cited in Boylan and Foley, p 111) [6]. Cairnes' *Character and Logical Method of Political Economy* (published in 1857) 'was never appreciated according to its merit' according to Schumpeter (p 824) [7]. Cairnes' book was focused on "the logical scheme of economic rationality commonly known as 'pure theory'" and his 'analysis of the nature of that schema was a real contribution' (ibid). However, it should be remembered that Cairnes' own or other works, or both, 'prove that he was far from believing this logical schema...comprises all our knowledge about economic affairs' (Schumpeter, p 824) [7].

Here I will briefly outline some of the implications of these nineteenth century debates for the development of an alternative institutional political-economy approach to telecommunications in the late twentieth century. It will indicate how Cairnes can be viewed as an important intellectual source for the development of alternative 'institutional' political-economy approaches to telecommunications.

2.2. Cairnes: on limits to the universalism of political economy

2.2.1. Cairnes' critique of dogma and universal formulae

Cairnes was one of the first influential economists within academe to recognise that laissez faire was a doctrine that had arisen out of specific social circumstances and industrial conditions in England. He argued that it was, therefore, neither applicable to nineteenth century Ireland nor anything like a universal scientific law or formula in itself. He criticised those who advanced political economy as a rigid set of principles that can or should be applied to every specific situation and policy problem: '....of late, the idea that all Political Economy is summed up in laissez faire has been much fostered by the utterances of some public men and writers who have acquired a certain reputation as political economists, chiefly, it would seem, through the pertinacity with which they have enforced this formula, insisting on its sufficiency, not merely in the domain of material interest, but over the whole range of human life.....I deny that economic doctrine is summed up in laissez faire; I contend that it has positive resources, and is efficacious to build up as well as to pull down.....' (Cairnes, pp 185–7) [8].

Cairnes' views on this matter were fully formulated by 1870 and it is clear from his writings that it was the Irish experience in the period from the famine to the first Land Act that provided the material and basis for his then quite revolutionary views – at least in relation to the prevailing official academic agenda (Deane et al [3]; Boylan and Foley [4]). As an academic political economist, Cairnes was anxious to defend political economy itself, as well as to effect improvements in the Irish land system in the midst of the political and social conditions in the decades after the famine.

His work suggested that it was precisely the kind of rigorous adherence to doctrinal economic belief, so characteristic of British politicians and officials in the nineteenth century, that had given political economy a bad name and had also contributed so much to Irish misery at that time. Indeed some commentators have suggested that, given the pressing socio-economic conditions of the time, it is as unsurprising as it is appropriate that an Irish economist should be foremost in exploding the malign myth of laissez faire and in advocating the necessity of state intervention to control and resolve the economic disorder generated by the system in Ireland (Boylan and Foley, p 117) [6]. As Cairnes himself put it: 'Laissez faire, freedom of contract, and phrases of like import have of late become somewhat of bugbears with a large number of people..... I hold it to be a pretentious sophism, destitute of foundation in nature and fact, and rapidly becoming an obstruction and nuisance in public affairs...' (Cairnes, p 191) [9].

Other key features of Cairnes' political economy are given in the annex to this chapter.

2.3. *The implications of this historical and theoretical excursus*

The political economy of Cairnes has relevance and lessons for those of us concerned with the complexities and other challenges involved in telecommunications policy in different socio-economic settings in Western Europe in the 1990s. Cairnes' writings and the prevailing orthodoxies which he sought to challenge and critique, suggest remarkably direct echoes and relevance to the issues confronting researchers in this very contemporary and specific arena. We too are confronted with the 'bugbears' of pervasive attempts to assert and impose rather simplistic and universal formulae based on a modern version of the kinds of laissez faire doctrines which Cairnes confronted in the last century.

In brief, I believe that Cairnes's ideas concerning the nature and scope of political economy and its relevance to practical policy issues contain some implicit and explicit insights of relevance today. Furthermore, they provide some useful insights for those of us concerned with developing more equitable approaches to telecommunications policy-making today. In particular, I believe that his contribution is particularly useful in relation to the following essential tenets of an institutional approach to telecommunications regulation.

1) It emphasises the fact that in the 1990s as much as the 1860s, economic 'laws' fall far below the scientific status and universal applicability that most of the contemporary proponents (high priests) of the discipline often suggest. Clearly, since the time of Adam Smith and Cairnes, economics has developed and adopted the mantle of a high degree of formal sophistication. But despite the great volume of complex formal theorising and econometric technique that now constitutes the field of economics, 'the ideological cart is often still in front of the theoretical horse', as Hodgson put it [10, p 14].

2) Secondly, Cairnes' contribution and example serves to remind us that the role and responsibilities of academic social-scientists is to critically evaluate, and if necessary

challenge, the prevailing fashions and orthodoxies of their day, whether in relation to policies for (tele)communications or any other sector.

3) Thirdly, Cairnes' work suggests that whatever the appropriateness or applicability of any economic theory or concept in any one national institutional context, it does not follow that it is equally appropriate in other, neighbouring national contexts. To repeat, he suggests that we should be wary of any reliance on alleged quick-fix and geographically 'universal' political-economy 'laws' or 'formula'.

4) There are no universal laws or rules of political economy (or 'formula' to use Cairnes' term) which are always relevant to the analysis of specific policy problems or which are equally applicable in all or different specific sectors, or both. In other words, any attempt at economic analysis must be very attentive to sectoral and other specificities. The blind and universal application of economic dogmas and formulae can become a real 'obstruction and nuisance in public affairs' (Cairnes, p 191) [9]. Here, in particular, we should remember Cairnes' injunction that many practical problems, involve aspects other than the purely 'economic'. For Cairnes, these other aspects may 'involve consequences so weighty as to turn the scale against purely economic solutions' (ibid). Cairnes suggested that these other (extra-economic) aspects include the 'artistic', the 'educational', and the 'moral'. But in keeping with this approach within our own time and concerns, we can readily add the terms 'communications networks' and 'infrastructure' here as well.

5) Cairnes stressed that many practical problems involve aspects other than the purely 'economic', and thus I would argue that this applies particularly to contemporary communications-policy issues. Here we must recognise that communications services and networks have been developed as increasingly important bases for all forms of social and political as well as cultural life in recent decades. In addition, in recognition of one of the key qualitative shifts in the economic fabric since the mid-nineteenth century, we can also add that the specificities of the 'technological' as well as the infrastructural aspects of telecommunications issues must be fully recognised and addressed as well. And if this requires stepping outside the conventional or universal frames of economic analysis, then so be it.

3. Developments in telecommunications technologies, markets and regulation in Europe

3.1. Introduction

The discussion of Cairnes' political-economy model in the previous section, and particularly his influence on subsequent thinkers, leads us naturally to favour an 'institutional' approach to economic and policy aspects of telecommunications problems. Cairnes did not label himself an 'institutionalist'. But his approach and emphases, as well as his influence on, or links with, the work of later writers such as Veblen and Keynes, locates him clearly within the political-economy tradition that we now label 'institutional' (for example, Hodgson) [10].

3.1.1. The institutional model

The institutional model of political economy is preferable within the communications sector because:

1) It is mindful and respectful of the specific socio-economic conditions and the particular institutional settings of different countries and regions.

2) It is attentive to the very specific and particular characteristics of different industries and sectors. In the case of the telecommunications sector, not the least important of these economic characteristics which this approach draws our attention to are the very peculiar existing and emerging forms and structures of competition within this sector.

3) It is also an approach which, like that of Cairnes, recognises that many practical problems involve aspects other than the purely 'economic'; and that these other aspects may involve consequences so weighty as to turn the scale against purely economic solutions (Cairnes) [9]. Thus, the selected institutional approach is one which refuses the 'expert idiocy' that often characterises the drawing of boundary lines within the social sciences.
4) It is able to directly address and take account of the role and implications of technological change in the telecommunications sector.
5) It recognises the existence and addresses questions of conflicting (as well as 'competing') interests and values in the communications sector – and indeed in any other sector.

Thus, in this section we will adopt the institutional model in order to examine recent developments in telecommunications-regulatory and other policies in Western Europe.

3.2 The neo-liberal agenda and hegemony

Since the latter half of the 1970s and the early 1980s, both academic and practical economic analyses and approaches to policy making have been dominated by neo-liberalism. As noted above, in the communications and other sectors, this shift has been marked by the growing (and now almost universal) dominance of policy tendencies labelled 'deregulation', 'liberalisation' and 'privatisation'. And within this shift, a growing role has been ascribed to market-based 'competition' as the key or primary source of regulation in communications markets (Preston and Grisold) [11].

The decline of the established 'public-service monopoly' approach to telecommunications in Western Europe in recent years, and the dominance or hegemony of the neo-liberal approach has been marked by a growing stress on the virtues and benefits of competition for all telecommunications users. (This was manifest, for example, in media debates in 1993 over the recent massive hikes in the charges for local telephone charges in Ireland.)

Apart from the predictable and often contradictory political and party point-scoring matches between the politicians, the media debate has been dominated by one theme: virtually all the 'pundits' and 'experts' who have pronounced on these issues in the media have blamed the tariff hikes on 'the lack of competition' in telecommunications and have called for more competition and privatisation of the public telecommunications operator (PTO) as the solution. In doing so they have created their own particular 'virtual reality' version of the telecommunications-service sector.

In applying economic text-book dogmas (or 'formula' to cite Cairnes), the majority of these so-called pundits have revealed a scant knowledge of the specificities of the telecommunications sector, its competitive structures or technological dynamics, or of the nature, role and direction of national and EU policies. (Not to mention the failure to address the specific set of factors which have led directly to the push for 'tariff-rebalancing' packages in the first place.)

The stress on increased competition as the core element of a new approach to telecommunications regulation is not confined to Ireland. It featured strongly in the US debates around 'deregulation' and 'disvestiture'. In the EU context, the vanguardist British approach to neo-liberal telecommunications regulation in the early 1980s was essentially based on beliefs and assumptions about the efficacy of competition as the main or only means of protecting 'the consumer interest against monopoly power' (Garnham) [12]. In Britain, these ideas continue to hold great sway – even if in this relatively large and well-favoured national telecommunications environment, the forms of competition which have actually emerged so far remain very limited and confined.

Now, in criticising conventional economists for being all too ready to apply their ready-made and universal formulae or assumptions about the efficacy of competition in this or any other sector, it is important to stress that I do not wish to propose the opposite trap of another universalism: one which ignores or denies the potentially significant changes that have occurred within the sector and their implications for increased competition and thus for new approaches to regulation.

3.3. *Forces leading to increased competitive pressures in the telecommunications-services sector*

In the relevant literature we can find three conceptually distinct sets of forces which are tending to push for the introduction of greater competition and new regulatory approaches in the telecommunications-services sector.

1) One is the prevailing international political fashion for neo-liberal policies already noted. But, for some commentators, this is not viewed as a primary or major cause in itself; it is more a reflection or response to other pressures.

2) A second set of forces is linked to the direction and implications of technological change. The majority of researchers, covering both the right and left of the political spectrum, have stressed the key role of technological change, and especially the 'convergence' between computing and telecommunications technologies and markets as a key force for policy change (for example, de Sola Pool [13]; Garnham [12]; Smythe [14]). Some have pointed to an expanding diversity of communications technologies (satellites, microwave, mobile, etc) which can provide functional alternatives to traditional wire-based telecommunications networks. Others still, have located telecommunications networks as one key site and component within a broader set of radical innovations which represent a major new technology system (information communications technology (ICT)) with pervasive applications and fundamental implications for all areas of socio-economic and political affairs.

3) The third set of forces identified in the literature is economic. One frequent factor stressed here is the role of large users in the form of multinational corporations (MNCs), especially those that are relatively high or intensive, or both, and sophisticated users of telecommunications services. These MNCs require an increasingly diverse range of communications facilities and services (in contrast to residential users and most small firms whose requirements still centre around POTS). These large users seek to reduce their telecommunications costs by bypassing the established PTO networks (or threatening to do so), for example, by demanding access to both their own or others' leased lines, or to alternative communications facilities (such as satellite and microwave) provided by new market entrants (for example, computer firms or other telecommunications companies); or sometimes by setting up their own 'in-house' systems. Hence the pronounced eagerness of this industrial interest-group to promote the opportunities for competition and liberalisation of telecommunications afforded by technical change. A second (overlapping) economic factor is related to the fact that demand and turnover in telecommunications services are growing relatively rapidly. Thus, they are viewed by investors and their advisors in the financial sector as a new 'high-tech' area for capitalist investment – and one that is potentially highly-profitable, if heavy users are targeted ('cream-skimming'). Such exciting investment opportunities can only be realised if competition and liberalisation policies allow.

3.3.1. *The very peculiar and limited forms of competition existing in 'de-regulated' telecommunications-services markets*

Even a cursory empirical examination of the actual structures of competition in telecommunications markets presents a very different picture to that suggested by the formulae and 'virtual realities' presented in conventional economic theory and textbooks. This is true even in the case of countries that are supposed to be the most 'liberalised' and open to competition.

The main forms of competition that exist in even the most liberalised and more prosperous national telecommunications markets are confined to the large business users market, ie, the area of 'cream-skimming'. The strongest areas of existing competition are the high-capacity routes and for large corporate users. Despite their often technically complex, sophisticated and diverse demands, this small but important groups of users have been the main recipients and beneficiaries of what 'competition' exists. The reasons for the concentration of competitive pressures and benefits on this sector are material and obvious. They relate to the shares of telecommunications usage held by the major companies (for example, one-third of British Telecommunications' (BT) revenues and one-half of its profits are derived from just 300 major-user companies (Garnham, p 146) [12]. Even the majority of business users, (the small and medium-sized enterprises) have not really shared in the promised benefits from competition in any significant way so far.

There is almost no competition in relation to local calls and the large group of users defined as residential customers. Even in Britain with its much-heralded and pioneering competitive regime (including unique incentives for the development of competition via cable-TV services), there is only token competition in the local loop (ie, for residential consumers). By 1992/93, BT still accounted for 98% of all residential telephone lines and calls in Britain (see Cave and Sharma, Chapter 1 of these proceedings).

In other words, what we find, even in the most competitive and 'liberalised' telecommunications regimes, is a highly biased and limited form of competition. And certainly, the empirical evidence demonstrates that not even in the most gung-ho neo-liberal regimes (for example, the UK) has telecommunications policy made any real progress towards realising the neo-liberal claim and ambition to introduce sufficient competition to make the industry largely self-regulating.

3.4. *Infrastructure: competing and complementary technologies/networks*

3.4.1. *Competing and complementary telecommunications technologies*

In research literature and policy debates, we frequently encounter the argument that new technologies are driving and demanding the introduction of competitive markets and changes in regulatory structures. A very common version of that argument suggests that thanks to developments in the underlying technology-field of microelectronics, we are now witnessing (i) a range of entirely new means and systems of electronic communications, and (ii) a 'convergence' or blurring of functional and market boundaries between existing communications sectors.

Proponents of this approach claim that these new-technology-based services render the inherited policy models and concepts of 'a monopoly public service', universal service, etc, outdated and irrelevant in an era of a multiplying choice of communications systems.

The most frequently mentioned and pertinent competing technologies in this context are:

a) The traditional 'fixed' or 'wire-based' telecommunications network and cable-TV networks.
b) Mobile cellular-telecommunications services.
c) Satellite (VSAT) services.
d) Microwave services (for example, MMDS/MDS).

The neo-liberal approach has recognised and tended to stress the tendencies towards convergence between these previously separate markets and technologies and has generally treated them as competitive and complementary communications systems. Especially in the larger pioneering 'liberalising' regimes, the neo-liberal approach of the state has proclaimed a 'hands-off' posture favouring competition and a determination of the technological winners and losers via the 'invisible hand of the market'.

This 'hands-off' posture is simply one more example of the 'virtual' reality constructed by neo-liberal ideology and dogma. For, despite the rhetoric or desires of neo-liberalism, the more concrete reality is that the state is inevitably and intricately bound up with the technology selection and competition processes involved. This is manifest, for example, in the frequently-encountered tendencies to 'encourage' fibre-optic networks. Such influences naturally flow from the whole range of direct and indirect state-expenditure decisions related to specific technology-development trajectories, whether in the form of military and procurement policies or explicit R&D programmes at the national/state or federal/EU levels. And of course, every nuance of regulatory policy impinges on the viability and relative prospects of any particular technology.

This neo-liberal concept is what lies behind the push for the open network architecture (ONA) approach in the USA (open network provision (ONP) in the EU). A recent review of the US case suggests that the idea of 'convergence'-based competition between the telecommunications, cable and computer industries has been simplistic, mistaken or at least overstated (Simon) [15].

For one thing, even in the relatively prosperous US context, the extent and forms of competition between the telecommunications and cable companies, in some respects, remains somewhat limited. Recently, some of the executives in the 'warring' cable and telecommunications camps have been advocating a seemingly radical notion: 'co-operation' (Simon, p 17)[15].

At the same time, in the US market, competition for broadband-type markets is not confined to convergence between cable and telephone companies. There are other suppliers ranging from interexchange carriers, competitive-access markets, the internal networks of big companies, and the dedicated networks of particular sectors. Here, the questions and problems of how to interconnect networks and break down services remain central to the concept of ONA and to the growing hype over the promise of the 'information super-highways'.

With regard to convergence, it has proved very difficult and costly to cross over from one sector to another. There have also been repeated failures of attempts to introduce new services on networks originally designed as single-use facilities. Part of the problem is the persistent low-levels of demand for many of the new services and products. Simon's analysis of US developments also suggests that a key problem is not so much technological or market crossover between the cable and telephone operators; rather it appears to be 'the very type of architecture that is best suited to meet not so much current needs .... but rather future needs' (Simon, p 20) [15]. Simon suggests that the fundamental question in relation to any shift towards broadband networks is whether or not there is any need for a global infrastructure. He suggests that in the USA, the 'uncertain context of shifting alliances' means that 'the chips will not necessarily fall on the side of homogenous networks' (Simon, p 21) [15].

3.4.2. But the 'infrastructure' concept of the telecommunications will not lie down and die

Even within the neo-liberal contexts of the UK and USA the ghost of another concept and idea about the future shape of the telecommunications infrastructure has not been totally banished. Rather it crops up with surprising frequency and often from unlikely sources. It is one that is particularly important and relevant for consideration in the context of small and peripheral economies with limited resources.

This refers to the concept of a 'common-carrier' approach to the broadband-communications network as a key infrastructure and 'core-technology' component of a major new-technology system for the future (Preston) [16] [17]. This idea cropped up in the Peacock Committee's report on the future of broadcasting in Britain. Even in the heyday of Thatcherite neo-liberalism in Britain of the late 1980s, the idea received serious consideration despite its antipathy to the prevailing dogmas (and the evident pressures towards ideological self-censorship operating within the policy-making elites of the time). However, the high priests of orthodoxy of the time did eventually manage to banish this ghost from the British agenda.

Many of those influenced by the neo-Schumpeterian approach and who view broadband as a key infrastructural component within a major new-technology system (ICT) are sceptical of the prospects of a pure market or competition model delivering the goods in an efficient or coherent form even in the larger and more prosperous economies (for example, Preston) [17].

However, within a neo-Schumpeterian approach that takes account of specific national socio-economic and institutional factors (for example, those considered in section 4) and their relation to this core new technology, the material scope and potential for investment in competing technologies and systems is particularly limited in the case of small and peripheral economies. Thus, the common-carrier approach seems particularly worthy of consideration in the case of these countries. This is particularly the case if one considers broadband (integrated services digital networks (ISDN) or some other variant) as the 'basic' service of the future, ie, one where access and geographical coverage requirements are as 'universal' as we take POTS to be today.

The infrastructural concept and approach has echoes in some of the recent discussions of the Gore/Clinton concept of a 'data-highways' and broadband as 'the super-highway' of the future. It is also one which finds active advocates (for example, Keegan) [18] in the UK policy-making debates of the early 1990s. Bolter [19] also discusses the different approaches to network-investment issues.

3.4.3. *Competition and complementarities between fixed and mobile telecommunications*

The shift to neo-liberalism has promoted not only the growth of many competing physical networks and technologies, but also the mushrooming of a host of competing concepts of technical and socio-technical and regulatory possibilities. Although these different concepts are often closely linked to particular industrial interests, they are delivered with varying degrees of transparency, opaqueness and hype.

Here, I want to deal with one influential stream of argument doing the rounds. It suggests that now, thanks to enhancements to capacity, quality and cost-structures, mobile (cellular), satellite and other wireless communications networks can replace fixed-wire telecommunications. This claim seems to be misleading or premature, at best.

Firstly, the 'fixed' local network is still generally cheaper than any wireless technologies in operation. Secondly, mobile systems will not be able to delivery the potentially large range of broadband multi-media and interactive services (for both residential and businesses users) that can be handled by fibre-based fixed networks. Thus to conclude, there is no substitute for the national fixed-wire system; it is an essential infrastructural investment form for the present and future (see also Adonis and Tait) [20].

3.5. *But do these changes imply the (universal) virtues of competition or monopoly in telecommunications supply?*

Neo-liberal theorists start out from the proposition that monopolies, and especially public monopolies, are inherently and universally detrimental to consumer welfare. Thus, they are predisposed to reject the idea of telecommunications as a 'natural monopoly' at the outset.

Since the late 1970s a number of researchers from both the left as well as the right sides of the political spectrum have argued that the notion of 'natural monopoly' in the telecommunications field has been undermined by technological and other developments (for example, Smythe [14]; de Sola Pool [13]). Smythe says "...the notion that telecommunications is a 'natural monopoly' in which the monopoly firm may serve and be regulated in a distinct, discrete market structure is now an out-of-date notion." (Smythe, pp 147) [14].

Others seem less dogmatic or certain about this. Some suggest that arguments about technological change and the growing diversity of services may be taken as powerful forces for change in the organisation and control of telecommunications. The importance of two developments are frequently mentioned here: (1) customer demands for service diversify, making it increasingly difficult, without some market mechanism, to respond efficiently to those demands; and (2) telecommunications and computing are not only converging but they are becoming ever more strategic across wide areas of economic and social life. For example, Garnham argues that these developments mean that 'the idea of one national monopoly controlling this development becomes both politically and managerially untenable.' [12](pp 150–51).

However, this does not necessarily mean that competition is an automatic or universal answer. For one thing, in the telecommunications sector, the case in favour of competition has to be demonstrated rather than assumed. This is because of the very limited, peculiar and distorted forms of competition that exist even in the most liberalised markets. Thus, given the limited real scope for competition and its inefficacy in network provision, it remains far from obvious as to how the development and expansion of the network and its associated services will be funded and controlled.

Here, I think it is necessary to consider a general theoretical point made long ago by Cairnes and one advanced by telecommunications researchers more recently: that in the final analysis, the case in favour or against competition may be as much a matter of faith and ideology as it is anything like 'scientific' economic analysis or universal panacea (Garnham, p 144) [12].

Furthermore, we also need to consider whether, in the contemporary telecommunications case, the answer might be not simply a matter of faith but also one of material constraints in the case of the smaller, less-developed economies in Europe (and elsewhere).

4. Implications for small and peripheral economies and the EU telecommunications context

4.1. Introduction

Here we will consider further, some of the issues considered in the previous section of this paper with a specific focus on the smaller and peripheral countries and regions (SPCR) of the EU. I will seek to investigate whether the scope for and role of competition might more be clear cut and less a question of a leap into 'faith and values' (section 3.5) in the case of small and peripheral economies compared with the larger and more prosperous ones.

In the case of these countries, it is necessary to consider whether a vigorous neo-liberal agenda might simply lead to:

i) At best, very limited and distorted forms of competition. Where accompanied by privatisation, the result might simply be one of a privatised monopoly simply replacing a public monopoly; a situation which has little justification even in neo-liberal writings.

ii) Resource-related difficulties in the development of effective regulatory structures, and a relative reduction in the effective powers or potential for regulation, or both, precisely in conditions where more rather than less regulation is required.

iii) Very strong and particular vulnerabilities of universal-service provision as a result of cream-skimming and bypass activities in the small countries and peripheral regions (SCPR).

iv) Real difficulties in terms of understanding how the development and expansion of the network (viewed as a core infrastructure in the present era) and its associated services will be funded, provided and managed.

4.2. *The EU's competition and liberalisation policy*

We have already touched upon some key elements of the EU's telecommunications policies in section 3 and their relation to the neo-liberal model. Since the mid-1980s, EU policies have attempted to create a single, integrated telecommunications market, and this has involved a significant attempt to formulate and apply 'universal' regulatory structures, rules and principles across all twelve member states.

4.3. *Special provisions for small and peripheral countries*

The main thrust of EU telecommunications policies is clearly based on essentially neo-liberal principles and directed at facilitating the creation of a single market and the creation of large and 'globally-competitive' industrial units and structures.

Within the neo-liberal perspective and approach, the special problems and issues related to the smaller, peripheral economies and regions, like other forms or dimensions of inequality, are essentially sidelined and marginal. The neo-liberal model asserts (or at least assumes) that 'competition' can be equated with 'competitiveness'. The ultimate logic of the neo-liberal perspective which informs EU economic and industrial policies is that the rising tide of prosperity will lift all boats in the long run (or, to reverse the metaphor, the so-called 'trickle-down' effect).

But, in practice, political pressures and realities have ensured that EU policies recognise and respond in some way to the threats posed for the less-developed regions by the push for a single market. The main EU telecommunications policies and documents (ie, apart from specific regional policies) make a number of concessions to the particular problems of the peripheral or less-developed economies of the EU and those with 'very small networks'.

In implementing it policies, the EU has taken some account of the differences and variations in the level of development between the regions (see Table 1). The EU policy-responses are also manifest in the structural and cohesion funds and the activities of the EU regional directorate. In practical terms, the EU has developed a two-pronged policy to stimulate a process of 'catching-up' based around: (i) financial contributions to telecommunications-infrastructure development; and (ii) institutional change. They indicate that, what some have termed the EU's 'obsession with competition' (Sousa, p 1) [21] is far from total or complete when we come to examine the full spectrum of EU telecommunications policies.

In the telecommunications sector, perhaps the most visible and practical manifestations of the EU's responses to the position of the smaller and less developed economies mainly take the form of the EU's structural and cohesion funds, and the STAR and Telematics programmes. These EU regional policies have clearly resulted in some shifts of investment funds to the less-developed economies for the purposes of network development and upgrading as well as the promotion of innovative services (for example, Sousa [21]; O'Siochru [22]). However, past experience suggests that explicit 'regional' policies, whether related to the telecommunications field or other sectors, tend at best to have only a marginal impact on the spatial distribution of investment, jobs and economic activity more generally. (This was true even in the heyday of Keynesian/social-democratic regional planning.)

Table 1. Summary of recent provisions for small countries in the main EU telecommunications-policy documents

1)	The EC notes: 'A realistic approach to further liberalisation must be followed, taking into account the need for adjustment in peripheral regions with less developed networks.' (OJC, C213/2)
2)	'The need to take into account the objective of Community cohesion in the light of the specific circumstances of peripheral regions.' (OJC, C213/2)
3)	'Major goals' have been identified as: (i) 'the working out of arrangements for suitable measures in relation to specific difficulties encountered by the peripheral regions with less developed networks. Such measures, as a complement to national funding, should where appropriate, and taking into account the priorities set at national level, make full use of appropriate Community support frameworks to assist network development and universal service in peripheral regions'; and, (ii) 'the taking into account by the Commission, in the preparation of the steps to implement the goals of this Resolution, of the specific situation of small networks.' (OJC, C213/3, 6.8.93)
4)	'Longer term' major goals are considered: 'the examination, prior to full liberalisation of all public voice telephony services, of progress on structural adjustments, in particular of tariffs, in those countries experiencing specific difficulties, in order to take account of the situation of the *peripheral regions* with less developed networks and of *very small networks*, including the fixing of additional transition periods, where justified.'
The July 1992 Council resolution stated that it gave its support to the Commission's proposals: 'to prepare before 1 January 1996 the necessary amendments to the Community regulatory framework in order to achieve liberalisation of all public voice telephony services by 1 January 1998.' But that 'In order to allow member states with less developed networks, ie, Spain, Ireland, Greece and Portugal to achieve the necessary structural adjustments, in particular of tariffs, these Member States are granted an additional transition period of up to five years.' Here the Council also noted 'the intention of the Commission to work closely with these Member States in order to achieve such adjustments as soon as possible and in the best possible way within the period. Very small networks, can where justified, be granted a period of up to two years.' (OJC, C213/3, 6.8.93)	

Note: This does not cover the policies of the Directorate General for regional affairs

The EU's recognition of the special problems of operators in small and peripheral economies and 'small networks' is also manifest in the main telecommunications-policy negotiations and documents (Table 1). In many ways, these are the arenae where the real stakes are located and where the major EU policy decisions affecting telecommunications developments and outcomes in the small and peripheral as well as the larger countries are determined. Of particular note here was the recent decision to grant a derogation or postponement of the 1998 deadline for the liberalisation of all public voice-telephony for 'up to five years' in the cases of Spain, Ireland, Greece and Portugal. Although this concession came as a surprise to some of those involved, developments since this policy was formulated suggest that it should not be read or interpreted as implying any relaxation in the feverish pace of liberalisation undertaken by the European Commission since 1987.

The most appropriate overall characterisation or summary of EU policies relating to telecommunications is their strong neo-liberal flavour and 'universal' scope and reach – increased competition in networks and services is possible and appropriate in all countries, large and small, core and peripheral. But this universalism is qualified, to a very limited extent, by a recognition of the specific socio-economic and technological conditions in the smaller, peripheral economies.

4.4. EU telecommunications policy and the specific problems of small countries

4.4.1. Introduction

What have been the main impacts and implications of EU policies for the present and future developments in the telecommunications-services sectors of the smaller countries and peripheral regions (SCPR) of the EU? Applying the approach and analysis outlined earlier (in sections 2 and 3), we wish to ascertain: what are the specific impacts and implications of EU policies for the small and peripheral economies of Western Europe; and how do EU policies match and 'fit' the particular socio-economic and institutional context of the small and peripheral economies of the Union?

The discussion that follows suggests that there are a number of specific and important sets of issues that EU telecommunications policies pose for the small and peripheral economies of Western Europe at the present time.

4.4.2. The limited scope and the peculiar forms of 'competition' in telecommunications in SCPRs

Earlier, it was noted that the organisation and market structures of the telecommunications sector indicate that is not possible to introduce sufficient competition in the telecommunications sector to make the industry largely self-regulating. Nowhere does it, or can it, approximate the liberal model of free and open competition, thus allowing the regulatory agencies to operate with anything like 'a light touch'. Even in the large telecommunications markets which appeared to pioneer and promote the neo-liberal approach to telecommunications regulation, we find that the actual extent and forms of competition are very limited. Competition is largely confined to the corporate-users end of the market, or the (primarily business) sub-markets concerned with newer services such as mobile and datacommunications. There is little or no competition at the level of 'the local loop' or for residential users.

The generally limited scope for competitive telecommunications-market structures to develop is particularly marked and evident in the case of small and peripheral countries (SPCs) and regions. We have already noted some of the peculiarities and distorted forms of actual and potential competition in the telecommunications-services sector. It is very clear that there is even less scope and potential for 'competition' in the case of residential users or in the local loop, and this is especially so in the case of the smaller and less developed economies. For the foreseeable future, monopoly is the name of the network game in these countries (even more clearly than in some of the larger and more prosperous economies which have pioneered the neo-liberal programme of telecommunications reform).

This can be indicated by a consideration of some basic comparative data related to socio-economic structures, institutional factors and the levels of 'communications development' in the two groups of countries:

A) In most of the small and peripheral economies (SPEs), we find not only relatively low levels of per capita GDP and final consumption, but in addition, these countries are also marked by both low absolute levels of population and very low measures of population per square kilometre. These data alone suggest a major funding obstacle related to the provision of two or more competitive networks in the case of SPEs (ie, a major hole on the demand-side of the models which assume the possibilities of competitive provision of services).

Applying the 'rational economic man' model of investment decision-making inherent in the neo-liberal approach to telecommunications planning, these characteristics clearly make the SPEs relatively unattractive locations; quite simply they are high-risk, low-return locations for investment by profit-orientated private-sector competitor telecommunications operators, compared with the larger, higher income and more densely-populated economies. EU structural and other 'regional' policies implicitly recognise and seek to compensate for some of these investment differentials. However, even cursory calculations of the relative sums involved suggest that these 'regional' funds do not amount to creating anything like an adequate 'level playing field' in this particular context.

B) The relatively low levels of 'final consumption' demand is compounded by relatively low levels of demand for telecommunications services due to the industrial structure of many small and peripheral countries. For example, countries like Greece and Portugal are still very dependent upon agricultural industries. Moreover, in countries like Greece, much of the agricultural sector is organised around very small-size farms and the population is highly dispersed spatially (for example, Michalis, p 6) [23].

Nor can capital cities in SPEs (such as Dublin or Athens) be compared with London or New York, either relatively or absolutely, as international business-service centres (ie, when it comes to the role of international financial and other business-services industries which are heavy users of all kinds of basic and advanced telecommunications services) (Constantelou) [24]. But when it comes to industrial structure and demand for telecommunications services, there are also some important differences between Ireland and some other SPEs (for example, Greece). In Ireland, for example, although agriculture remains an important 'indigenous industry', the 'high-tech' end of the manufacturing sector is dominated by MNCs with very heavy and sophisticated telecommunications-service demands and use.

C) The potential brakes on competitive entry associated with the socio-economic specificities identified above (for example, relatively low levels of potential demand, low population density, etc), are further compounded in some SPEs by the prevalence of mountainous terrain (for example, 80% of Greece is covered by mountains) and other costs imposed by physical geography.

4.4.3. Particular threats of 'bypass' and 'cream-skimming'

There is some evidence to suggest that in most, if not all of the SPEs, it is probable that the current and future impacts of 'by-pass' and 'cream-skimming' will be, in relative terms at least, much more severe than in the larger and more prosperous economies.

For several reasons, most notably the lack of economies of scale and scope, technological competence and sophistication, and marketing power and global reach, the telecommunications operators based in the SPCs (whether privatised or not) will not be able to compete with the giants based in the larger and more prosperous economies. These three factors are very important in the only really competitive segment of the contemporary telecommunications-services market-place: the provision of increasingly sophisticated, heterogenous and technology-dependent services to large corporations.

For example, the annual turnover of British Telecommunications is fourteen times that of the Irish PTO, Telecom Eireann (TE). Indeed, BT's annual profits are more than double the total annual revenues of TE [25]. In relative terms, the telecommunications operators of the SPCs are much more vulnerable to competition, especially in the form of 'cream-skimming' than are the operators based in the larger and more prosperous economies. As soon as competition is officially permitted, the large overseas operators, with their vastly superior technical and marketing resources, will undoubtedly prove to be the most attractive proposition for the large and medium-sized firms within SPCs, especially those with international telecommunications-services requirements.

Even before the official licensing of competition, the revenues of some PTOs based in the SPCs were under pressure from sophisticated 'grey' (if not directly illegal) forms of bypass and competition. The grey or illegal bypass issue is compounded in some SPCs by the small scale of regulatory resources and expertise available to effectively monitor, implement and interpret existing national and EU regulations.

4.4.4. *Tariff 'rebalancing' and residential users*

One direct and important impact of EU telecommunications policies in some SPCs has been recent and significant hikes in the charges for local calls. For example, the May 1993 tariff rebalancing in Ireland has been the source of some front-page newspaper stories and led to one of the most intense political rows during 1993 (Preston) [26].

As in other countries, this 'rebalancing' is based on a very specific approach to the allocation of costs between the various components of the telecommunications networks and markets. As in other countries, it is a regulatory shift which reflects the interests and conceptions of large corporate users and, in economic terms, it involves a redistribution of income from residential users to the corporate sector.

Again, there is some evidence to suggest that these kinds of shifts and impacts have been more severe in the SPCs than in the larger and more prosperous economies. Certainly, if the earlier discussion of the particular vulnerabilities of operators in SPCs to competitive pressures, and to the problems of bypass, is accurate, then this has important implications for the future levels and structures of tariffs in SPCs.

4.4.5. *EU policy and infrastructure development issues*

A number of researchers have indicated the dangers inherent in a gung-ho pro-competition policy in terms of the investment in, and maintenance of, a coherent universal infrastructure for the future. As suggested earlier (section 3.4), this concern has been linked to the issue of ensuring a continuation of 'universal service' (geographical, affordable, etc) in the future as the norms or standards of 'basic' services shifts to ISDN and other broadband facilities. The Commission of the European Communities has promised a Green Paper on telecommunications-infrastructure issues and cable-TV networks by 1 January 1995. Apparently this will address issues of competition and its implications for geographical and other aspects of equal access and universal service.

Clearly it is too early to judge the EU's overall policy position on this very important matter. But if the analysis so far in this section is accurate, then the financing and further development of the infrastructure is going to be a particularly challenging problem for the smaller and peripheral economies. For example, even today before the full brunt of EU competition policy is fully felt in SPCs like Ireland, it has been estimated that even if TE were to undertake all the investment in the infrastructure that could be justified on a commercial basis, this would fall short of the levels that would be required to meet wider economic, social and political goals (NERA) [27].

At the very least, tackling this issue will probably require a certain 're-balancing' of emphasis from the heavily industrial and economic orientation of EU telecommunications policies in more recent years. It will probably require a shift towards concerns with the social, cultural and indeed traditional political dimensions of telecommunications.

This is one very important area where independent and comparative research could play a major role in monitoring and shaping future developments, especially as universal and affordable access to telecommunications is likely to become an increasingly important prerequisite for, and influence on, participation in social, economic, and cultural activities, for the reasons just outlined above.

4.4.6. *The relatively high costs and burdens of 're-regulation' in SCPRs*

Neo-liberals place great faith in the efficacy of the 'hidden hand' of competition and the market to allocate resources and protect the public interest. But whatever the real or imaginary virtues of market-competition in other sectors, the empirical evidence clearly indicates that this cannot apply in the case of the telecommunications sector. The very peculiar and limited forms of 'competition' prevailing, and likely to prevail, in the telecommunications market-place mean that any effective and meaningful regulation will have to be provided via a public or state regulatory body. As others have stressed before, the result is that we are confronted with a situation more properly described as 're-regulation' rather than one of 'de-regulation'.

The experience of the USA and Britain indicates that the development of new telecommunications regimes at national and EU level will involve a shift towards a more rather than less active role on the part of state regulators. The state regulatory agencies will have to undertake much more research and monitoring of market developments and play a more active regulatory role to ensure an 'even playing field' amongst the competing actors and to ensure that public or universal-service goals are being met.

This kind of regulatory role is necessarily expensive and resource-hungry in terms of the provision of expert personnel equipped with an appropriate mix of technical, legal, economic and policy-making knowledge in this area. In both large and small countries the neo-liberal regulatory model involves the development of sophisticated new concepts and the skills required to monitor the regulatory system (see Cave and Sharma, Chapter 1 of these proceedings). Small countries' governing elites are then faced with a choice between heavy expenditures on these regulatory resources or of opting for some less than satisfactory 'light-hand' approach to inherently oligopolistic (if not simply private monopoly) markets.

Any meaningful attempt at effective regulatory action in the telecommunications field requires a considerable level and mix of technical, managerial, socio-economic and 'political' expertise which is independent of the major interests involved in the particular national market. Such a combination of skill and resources is particularly expensive and burdensome in the case of SCPRs compared with their larger and more developed neighbours.

Indeed, it is one which proved to be a problem even in the larger countries that pioneered and embraced the neo-liberal model of telecommunications relatively early. But even in the USA with its relatively well-funded and highly-resourced Federal Communications Commission (FCC) (at least compared with other countries), it is often claimed that its range and depth of expertise in certain disputed areas has been clearly lagging behind that available to the contestants in some disputes.

Unlike the USA, or some of the larger EU states such as the UK, the small and peripheral economies do not possess a tradition or culture of utilities-regulation – nor the whole 'penumbra of expert economists, lawyers and academics and its tradition of public-interest debate and action around regulatory issues' (Garnham, p 141) [12]. In fact, the costs and implications of effective regulation under monopolistic or oligopolistic competition have not been fully recognised or addressed in neo-liberal writings or in EU documentation.

This resource-problem has led some countries to develop very particular approaches in an attempt to accommodate the requirements of EU policy, for example, in relation to audiotext services (Preston) [28] The resource-problem in SPCs is such that, in the case of Ireland, the telecommunications policy unit within the Department of Communications consists of only ten people. This small group is charged with responsibility for attempting to help develop, formulate and negotiate policy changes at both the national and EU levels (and to engage in consultations with the various interests at stake in the spate of policy changes). It is a tiny body of people and expertise compared to the resources of Oftel in the UK and especially with the FCC in the USA.

The stakes involved here are more than simply those of 'administrative failure': the resource-constraints that block the funding of 'refereeing' resources that are supposed to be required for the implementation of the neo-liberal model of regulation. There are two dimensions here:

1) Compared with the better-endowed agencies in larger economies, the national regulatory bodies in the SPCs are less able to alert, consult with and consider the views of

national industry in relation to EU telecommunications policy changes. This suggests that firms and industries in the SPCs (especially small-scale, and indigenous industry) are more likely to have to react and respond to EU policies (as opposed to having some input, however indirect, into their formulation) compared with their counterparts elsewhere. This appears to be less of a problem in the case of branches of MNCs located in the SPCs (as these generally have their own internal corporate channels which alert them to impending changes which may affect their competitive position, etc). This can be defined as one other aspect or dimension of the 'democratic deficit' and the relative exclusion of the interests of small entities that currently surrounds EU policy-making in this and other sectors. It is a reflection and extension of the market disadvantages already confronting small firms or entities into the sphere of 'political competition' in relation to the telecommunications and ICT sectors.

2) The role and influence of small and peripheral economies in shaping the form and content of EU telecommunications legislation may be even less than their already low shares of the official distribution of votes in the various fora would suggest. In Ireland, the paucity of regulatory resources is also evident in the radio-broadcasting field. Here, we have seen that the introduction of competition in radio broadcasting in 1988 has not been accompanied by anything approaching a meaningful attempt at monitoring the performance or compliance of the various licensees. This experience is not particularly encouraging for the prospects or requirements for effective 'neo-liberal' forms of regulation in the potentially more-complex field of telecommunications regulation. The broadcasting case shows us that small countries like Ireland lack the resources to develop even the most basic forms of effective regulation which are supposed to accompany the neo-liberal regulatory regime.

4.4.7. Technical standards and related policy issues

The small and peripheral economies are also disadvantaged by similar resource-constraints in relation to the processes of standards-setting and related policy issues. They are able to play a much less active role and make lower inputs into these industrially and commercially-significant proceedings compared with their larger neighbours. Essentially, they end up more as passive recipients of technical and related standards which serve to shape and influence the design and specification of new telecommunications and other technologies.

4.4.8. EU telecommunications policy and its role and impact on residential users

Although numerically the most significant group, residential users are excluded from any real voice or influence in national or EU telecommunications policy-making. They are also the group whose interests have been most negatively affected by the 'impacts' of recent policy changes (Preston) [26] [29].

When related to the empirical evidence (for example, that related to tariff charges), the neo-liberal model's claims and assumptions that increasing competition inevitably leads to enhanced consumer welfare and benefits are clearly misleading. At best, they amount to no more than wishful thinking based on erroneous assumptions about the scope and forms of potential competition in this sector (reflecting the imaginary or 'virtual' rather than concrete realities of the sector) (Preston) [26]. At worst, they amount to a legitimating device for a significant redistribution of economic benefits towards one interest group (large business-users) and increasing the burden of telecommunications costs for another, 'the silent majority' of consumers of services – residential telephone-users.

The recent Irish experience of massive hikes in tariffs for local calls together with the trends and considerations outlined earlier in section 4, suggest that this is likely to be an area of particularly important common-concern for policy researchers in small and peripheral economies in the coming years.

4.5. Relation to the un-employment crisis

The 1992 EC policy documents explicitly note and mention that telecommunications policy should take note of the 'overall employment situation' both within and beyond the telecommunications sector (ie, the unemployment crisis). In Ireland, where unemployment figures total more than 300,000, all the major political parties have agreed that tackling unemployment is a major political priority. So an important, if rarely addressed, question is the following: how do telecommunications-policy options directly support and supplement job-creation priorities and policies.

For one thing, national and EU liberalisation policies are putting big financial and other pressures on PTOs to shed labour. In all of the SPEs, the PTOs are very significant direct sources of employment, especially of skilled manual labour. Already, the new policies, combined with technological changes, have led to significant staff reductions in recent years. In Ireland, TE has shed over 5,000 jobs in the past decade and similar job losses have been encountered in other SPEs such as Greece (Constantelou) [24].

Current neo-liberal policies will, if unaltered, inevitably lead to higher levels of aggregate unemployment via major reductions in the direct-employment role of the PTOs in the immediate future. For the reasons outlined earlier in this section, the negative employment impacts of these policies might be felt particularly severely in the case of the small and peripheral economies. For example, unless current trends and policies are reversed, it is quite likely that they could lead to the loss of a further 5,000 or more jobs (or almost half the current total number of jobs) in Telecom Eireann over the next five years as pressures of bypass and cream-skimming increase.

The big question is where and when will new jobs come, especially in the current sluggish labour market and economic climate throughout Europe and the OECD countries. By their very nature, the new entrants, with their narrow focus on large-scale users, will not provide any more than a small percentage of those jobs lost in the PTOs.

Mainstream neo-liberal thinking often ignores (denies) that unemployment is itself a major problem or concern. And when it does recognise or address the issue, it offers only a very limited grasp of the dimensions and realities of the problem and a very restricted and technical discussion of policy options which abstract from the concrete realities and dimensions of the problem (for example, see Brittan) [30].

Clearly some new, urgent and alternative forms of policy initiatives are required if these growing unemployment problems are to be tackled and remedied. Here, I would like to suggest again that collaborative and comparative research could play an important role both in mapping the scale and dimensions of the problem and in supporting the development of more appropriate policies.

I believe that this kind of research is urgently needed to examine the potentially negative implications of these developments (technological and regulatory changes) on the already severe unemployment crisis of the small and peripheral economies. Such research could also address the scope and possibilities for developing innovative initiatives and responses to these policy and technological pressures. Empirically-informed collaborative and comparative research could play a role in supporting new initiatives on the part of management and unions within the smaller networks and PTOs as well as in the public-policy field.

Here, there may be a useful role for comparative and cooperative research in SCPRs that focuses on new forms of incentives and structures to promote internal (product as well as process) innovation within PTOs. Another focus might fall on the potential role of the large PTOs as 'umbrella' or support structures for small innovative companies involved in developing new ICT-based products for domestic and overseas markets.

The analysis presented above suggests that alternative and more realistic models of the appropriate organisation and structure of PTOs in the smaller and peripheral economies are urgently required. In this context, collaborative research might also involve exploring positive and forward-looking (rather than defensive) new approaches to public service in the telecommunications sector. One possible option here might involve new forms of employee share-ownership and other internal incentive schemes within the PTOs to promote universal service and internal innovation. From the point of view of the majority of both workers and consumers, this could provide viable alternatives to the assumed and misleading 'universal' virtues of the privatisation model. Another possible element of this alternative approach might involve new forms of cooperation and even 'joint-ventures' between the PTOs of the smaller EU countries (to achieve economies of scale and scope in the development of new innovations, while retaining some degree of local control and enhancing economic benefits).

These examples are only indicative of the important issues that could be addressed by collaborative and cooperative research, focused on the employment and economic issues facing SPEs associated with the current restructuring in the telecommunications sector.

5. Some conclusions and implications

In section 2 of this paper I examined the work of an important but neglected Irish political economist, J E Cairnes, concerning the nature and status of economic concepts in relation to policy-making issues. I identified a number of insights and implications drawn from Cairnes' work for an institutionalist approach to the study of contemporary telecommunications-policy issues.

Firstly, it emphasised the role of critical academic research in questioning and, if necessary, going against the grain of current orthodoxies and dogmas – no matter how fashionable, dominant and hegemonic they appear at particular times. Secondly, it underlined the need for economic and policy research to be wary of 'universal' political-economic laws or generalisations and to be attentive to the specificities (sectoral, institutional, technological) of any particular research problem or area. Thirdly, it demonstrated that we should recognise and make explicit that, contrary to the prevailing 'technocratic' orthodoxies, much economic and policy analysis cannot avoid questions of clashing economic interests and political values.

In section 3, in keeping with these principles, I adopted and applied an 'institutional' political-economy approach to examine the features and implications of important economic, technological and policy developments in the telecommunications sector, particularly those related to the increasing role of the EU in the telecommunications field. I suggested that the assumptions and claims underlying the prevailing neo-liberal approach to policy-making in the telecommunications field seemed very far removed from the empirical developments. Despite the formal sophistication of the neo-liberal model, it constructs an imaginary or 'virtual' model of reality which ill-matches the empirical or concrete realities and developments in the sector as observed by this researcher.

The kind of institutional approach sketched here, allows one to move beyond the 'virtual reality' of neo-liberalism to a more realistic analysis – one that is informed by the concrete and material technological and sectoral conditions in telecommunications. The analysis points to particular problems about the scope for, and implications of, network competition in the case of telecommunications in the small and peripheral economies of Western Europe.

I demonstrated that, despite the much-heralded processes of convergence and competitive pressures arising from technological change and the growing heterogeneity of telecommunications services, the telecommunications sector was still characterised by very specific, limited and peculiar forms of competition. This continued to be valid in the case of even the most 'liberalised' telecommunications regimes. The 'common carrier' and infrastructural concepts and approach to advanced telecommunications networks has continued to pose challenges for the competing technologies and networks-approach favoured by neo-liberal thinking. This is manifested in the recent 'super-highways' initiative by Vice President Gore in the USA and it still crops up in UK policy debates.

The particular vulnerabilities of PTOs in small and peripheral economies (for example, their exposure to bypass and cream-skimming pressures, etc) pose very important challenges to the prevailing EU and national liberalisation-policy regimes, especially in relation to the funding of expanded geographical or universal access to broadband services.

In section 4, I moved on to analyse the implications of current EU (and national) liberalisation and single-market policies in the telecommunications field, with a particular focus on the implications for the small and peripheral economies in Western and Central Europe.

I discussed some of the features of recent EU telecommunications policy, and the special provisions for small economies and the peripheral regions. I argued that there were grounds to expect that the potential for competition, in the local loop especially, was even more limited in scope in the case of the small and peripheral economies. Ultimately, for the majority of telecommunications users, the present neo-liberal approach to 'liberalisation' and its privatisation policies is likely to merely result in some form of 'privatised monopoly' replacing a public monopoly, at least in the foreseeable future. This empirical outcome is one that is not only out of keeping with the claims of the neo-liberal approach to policy, but it is one that finds little justification or advantage within that model.

In the telecommunications-market segment where the selective forms of competition are most pronounced, the large business-users segment, the PTOs in the small and peripheral economies will be particularly vulnerable to competitive pressures. Indeed, if present policy directions continue, the PTOs in the small and peripheral regions will not be playing on anything like a 'level playing field' within a unified EU telecommunications market. They are likely to be particularly vulnerable to market-share losses (and hence revenue losses) as a result of the selective forms of competition (for example, lack of economies of scale, the greater scope for bypass, the ability of large private operators to 'cream-skim the profitable large business users in any small country).

The paper also identified and discussed other aspects of the specific problems and pressures posed for the telecommunications-services sectors of the small and less-developed economies in the light of current EU (and national) liberalisation policies. Section 4.5 sought to relate these telecommunications-policy issues and alternatives to the important policy issue of the unemployment crisis – something that has been identified as a major political priority in SPEs such as Ireland.

Clearly, the discussion and analysis presented in this paper suggests that the neo-liberal approach to telecommunications regulation which currently informs both national and EU telecommunications-policy approaches, is problematic, both in theory and practice.

The issues and evidence identified in my research suggests that this approach to EU and national policy-making has some particularly negative implications for the future development of telecommunications services and related industries in the small and peripheral economies within the EU. Thus, in response to these developments, this paper also outlined a number of potential topics and policy issues related to the SPEs, which might usefully (and urgently) be addressed by a network or team of researchers with interests in these matters.

This paper is very much a tentative attempt to sketch out some of the issues related to the implications of current technological and policy shifts for the telecommunications sector in the small and peripheral economies of Western and Central Europe. I am fully aware that I have been rather provocative and assertive in the case of some of the complex issues addressed in parts of this paper. But I believe that this is necessary to stimulate the necessary debate around these challenging issues.

Annex

Other key features of Cairnes' political economy

Cairnes' 1870 inaugural lecture as Chair of Political Economy in University College, London, presented the most accessible summary of his key ideas on these matters. It represented a very powerful attack on the laissez faire doctrine, or at least the belief and practice that it was a universal guide to action. He also challenged the notion that the overall general or public interest (welfare) is ensured by each individual pursuing his or her own particular interests. Thus, Cairnes presented a very much less rosy and optimistic picture of the unlimited virtues of the free competitive market, at least compared with Adam Smith and his nineteenth century followers within the 'respectable' academic tradition of political economy.

1) Cairnes was very critical of the assumed overlap between the pursuit of individual interest and that of the general interest, and of the benefits of the invisible hand (which one finds in most classical and neo-classical economic thought). He said that 'nothing is easier than to show that people follow their interest, in the sense in which they understand their interest. But between this and following their interest in the sense in which it is coincident with that of other people, a chasm yawns. This chasm in the argument of the laissez faire school has never been bridged. The advocates of the doctrine shut their eyes and leap over it.' (Cairnes, p 190) [9].

2) Cairnes was clearly aware of the necessity and scope for policies and programmes for social and industrial reform. Cairnes declared: 'It follows that there is no security that the economic phenomena of society, as at present constituted, will always arrange themselves spontaneously in the way which is most for the common good.' In other words, he argued, 'laissez faire falls to the ground as a scientific doctrine' (Cairnes, extract, p 191) [9]. Cairnes did not believe either in 'the opposite principle of state control' either. He said that 'as a practical rule, I hold laissez faire to be incomparably the safer guide' but it is only 'a practical rule and not a doctrine of science'. He said that like 'most other sound practical rules', this rule too is ' liable to numerous exceptions'. He argued that it is 'above all, a rule which must never for a moment be allowed to stand in the way of the candid consideration of any promising proposal of social or industrial reform.' (Cairnes, pp 190–191) [9].

3) Unlike the Smithian tradition, Cairnes fully recognised and emphasised that class inequalities are significant and major features of capitalist industrialism. He rejected the (essentially) Smithian argument that inequalities do not matter as the rising tide of capitalist industrialism will lift all boats to a higher level than in any other system. He explicitly recognised that class interests are not identical and urged that this must be taken into account.

4) Cairnes also argued that many practical problems involve aspects other than the purely 'economic'; these other aspects include (especially for us) the 'artistic', the 'educational' and the 'moral'. For Cairnes, these other aspects may 'involve consequences so weighty as to turn the scale against purely economic solutions.'

5) Cairnes passionately rejected several other aspects of the dogmatic and rigid approaches to political economic analysis. He explained that economic beliefs are culturally derived and that therefore they should not be clung to in the face of circumstances that contradict them. In saying this, Cairnes undermined laissez faire as though it were a law of nature and anticipated the policies of state intervention that were increasingly applied in Ireland and were later, with J M Keynes, to become the governing feature of British and Western economic theory in the mid-twentieth century. (Deane et al, p 184) [3]. (But it should be noted that there remain some tensions/problems in his treatment of the scientific/neutralist character of political economy, especially when viewed in terms of present-day epistemology, theories of knowledge, etc.)

Notes and references

[1] K. Marx, Economic and Philosophical Manuscripts (1844). Marx Early Writings. Penguin, Harmondsworth, 1975.

[2] Kevin Whelan, (review of Boylan and Foley, 1992, see [4]), *The Irish Reporter*, Issue Six, 1992.

[3] Seamus Deane et al (Eds), The Field Day Anthology of Irish Writing. 3 Vols (see especially Vol. 2). Field Day Publications, Derry, 1991.

[4] T. A. Boylan and T. P. Foley, Political Economy and Colonial Ireland: the propagation and ideological function of economic discourse in the nineteenth century. Routledge, London, 1992.

[5] Antoin E. Murphy (Ed.), Economists and the Irish Economy: from the eighteenth century to the present day. Irish Academic Press, Black Rock, Dublin, 1984.

[6] T. A. Boylan and T. P. Foley, John Elliott Cairnes, John Stuart Mill and Ireland: some problems for political economy. In: A. E. Murphy (Ed.), Economists and the Irish Economy: from the eighteenth century to the present day. Irish Academic Press, Blackrock, Dublin, 1984.

[7] J. Schumpeter, History of Economic Analysis. Allen and Unwin, London, 1982.

[8] John Elliott Cairnes, Essays in Political Economy (1873). In: Seamus Deane et al (Eds.), The Field Day Anthology of Irish Writing, Vol.2. Field Day Publications, Derry, 1991.

[9] J. E. Cairnes, Political Economy and Laissez-faire. (Introductory lecture delivered after his appointment as Professor of Political Economy at University College, London in 1870). In: S. Deane et al (Eds), The Field Day Anthology of Irish Writing, Vol. 2. Field Day Publications, Derry, 1991. (Page numbers in this paper refer to this version of the text.)

[10] G. M. Hodgson, Economics and Institutions: a manifesto for a modern institutional economics. Blackwell, Oxford, 1988.

[11] Paschal Preston and Andrea Grisold, Unpacking the Concept of Competition in Media Policy Making – the case of Austria and Ireland. Paper presented at the 'Europe in Turmoil' conference of the International Association for Mass Communication Research, Dublin, June 1993.

[12] N. Garnham, Telecommunications Policy in the United Kingdom. In: N. Garnham, Capitalism and Communication: global culture and the economics of information. Sage, London, 1990.

[13] I. de Sola Pool, Technologies of Freedom. The Belknap Press of Harvard University Press, Cambridge, Mass., 1983.

[14] D. Smythe, Dependency Road: communications, capitalism, consciousness and Canada. Ablex, Norwood, New Jersey, 1981.

[15] Jean Paul Simon, Broadband Networks: will there be a cable war? Paper given at Programme on Information and Communications Technologies workshop, Kenilworth, England, 1993.

[16] Paschal Preston, The Future of the UK Communications Infrastructure: some issues related to competing technologies. Paper presented to Confederation of Information and Communication Industries Working Party on the Future Communications Infrastructure, London, November 1987.

[17] Paschal Preston, State and Market Regulation of Major New Technology Systems: the case of electrical technology. Paper presented at Centre for Communication and Information Studies (CCIS) conference on 'Planning the Telecommunication Infrastructure'. CCIS, London, 1990.

[18] Victor Keegan, *The Guardian* (UK newspaper), 22.5.1993.

[19] W. Bolter, Accelerating Network Investment, *Centre for International Research on Communication and Information Technology Newsletter*, 5(7), 1993, pp 3–5.

[20] Andrew Adonis and Nikki Tait, Mobiles break into the big time, *Financial Times*, 19.8.1993.

[21] Helena Sousa, The Impact of EC Communications Policy in Portugal. Paper presented at the 'Europe in Turmoil' conference of the International Association for Mass Communication Research, Dublin, June 1993.

[22] S. O'Siochru, The EC's Telecommunications Policy and Less Developed Regions. Paper presented at the 'Europe in Turmoil' conference of the International Association for Mass Communication Research, Dublin, June 1993.

[23] Maria Michalis, Towards a New Telecommunications Era in Greece. Paper presented at the 'Europe in Turmoil' conference of the International Association for Mass Communication Research, Dublin, June 1993.

[24] Natasha Constantelou, The Greek Road to Reforms in Telecommunications: how far can the UK experience be useful? Paper presented at the 'Europe in Turmoil' conference of the International Association for Mass Communication Research, Dublin, June 1993.

[25] It was not an accident nor merely a question of ideology that the Thatcher government pushed for the opening up of EC telecommunications markets in the wake of BT's privatisation. It's active promotion of a pro-competition and free-trade regime in telecommunications markets elsewhere was clearly based to a great extent on perceptions of BT's comparative advantages compared with other PTOs, the

economic advantages and benefits that were perceived to flow to BT, and the British national economy (and government revenues) as a result. This is made quite explicit in a number of British policy documents of the 1980s.

[26] Paschal Preston, The Role of Users in Irish Telecommunications Policy Making. Paper given at Academic Network on European Telecommunications workshop, Paris, May 1993.

[27] National Economic Research Associates, Telecommunications and Economic Development in Ireland. Consultant's report, Dublin, February 1993 (unpublished).

[28] P. Preston, Development and Diffusion of Audiotex in Ireland. Paper given at Academic Network on European Telecommunications workshop, Vienna, May 1992.

[29] Paschal Preston, Irish Telecommunications in Transition, *Teaglaim 1*, 1993.

[30] Samuel Brittan, Presidential address to the British Association, August 1993 (unpublished).

List of contributors

Jacques Arlandis	IDATE, Montpellier, France
Yannis D. Caloghirou	National Technical University of Athens, Greece
Martin Cave	Department of Economics, Brunel University, UK
Gerhard Fuchs	Akademie für Technikfolgenabschätzung in Baden-Württemberg, Stuttgart, Germany
Anders Henten	Institute of Social Sciences, Technical University of Denmark
Herbert Kubicek	Telecommunications Research Group, University of Bremen, Germany
Steen Ottosen	Telicon, Denmark
Paschal Preston	COMTEC, Dublin City University, Ireland
Yogesh Sharma	Department of Economics, Brunel University, UK
Frank Stöckler	Wissenschaftliches Institut für Kommunikationsdienste, Bad Honnef, Germany
Thea Weijers	TNO Centre for Technology and Policy Studies, Apeldoorn, the Netherlands